Ancient Aliens in the Bible

Enoch, Giants, and Otherworldly Visitors in Jewish & Christian Lore

A Modern Translation

Adapted for the Contemporary Reader

Various Ancient Writers

Translated by Tim Zengerink

Table of Contents

Preface - Message to the Reader

What If You Could Help Rebuild the Greatest Library in Human History?

Thousands of years ago, the Library of Alexandria stood as the crown jewel of human achievement — a sanctuary where the collected wisdom of every known civilization was gathered, preserved, and shared freely.

And then, it was lost.

Through fire, conquest, and the slow erosion of time, humanity lost not just books — but ideas, dreams, discoveries, and stories that could have changed the world forever.

Today, the Library of Alexandria lives again — and you are invited to be a part of its restoration.

Our mission is simple yet profound:

To rebuild the greatest library the world has ever known, and to translate all timeless works into every language and dialect, so that no seeker of knowledge is ever left behind again.

By joining our movement to rebuild the modern Library of Alexandria, you become part of an unprecedented mission:

- **Unlimited Access to the Greatest Audiobooks & eBooks Ever Written:**

  Instantly explore thousands of legendary works—Plato, Shakespeare, Jane Austen, Leo Tolstoy, and countless more. All instantly available to read or listen, placing a complete literary universe at your fingertips.

- **Beautiful Paperback & Deluxe Editions at Printing Cost**

Own any title as an elegant paperback, deluxe hardcover, or stunning collectible boxset—offered to you at true printing cost, delivered straight to your door. Build your personal Library of Alexandria, crafted for beauty, built for durability, and worthy of proud display.

- **Fresh Translations for Modern Readers—in Every Language & Dialect**

Enjoy timeless masterpieces reimagined in clear, contemporary language—no more outdated phrases or obscure references. Alongside the original versions, we're tirelessly translating these classics into every language and dialect imaginable, ensuring accessibility and understanding across cultures and generations.

- **Join a Global Renaissance of Literature & Knowledge**

You directly support expanding our library, publishing deluxe editions at true cost, translating works into all global languages, and bringing humanity's greatest stories to people everywhere. By joining today, you're not just preserving a legacy of masterpieces; you set in motion a powerful wave of literary accessibility.

Become a Torchbearer of Knowledge.

Join us for free now at **LibraryofAlexandria.com**

Together, we will ensure that the light of human wisdom never fades again.

With gratitude and a shared love of knowledge,
The Modern Library of Alexandria Team

Visit:

www.libraryofalexandria.com

Or scan the code below:

Introduction

Reframing the Watchers:
Myth, Metaphor, or Memory?

In the modern era, few biblical mysteries have captured the imagination like the story of the Watchers—the enigmatic beings of 1 Enoch who descended from the heavens, mated with human women, and gave birth to a race of giants known as the Nephilim. These tales, preserved in ancient Jewish and Christian apocryphal writings, straddle the line between religious myth, esoteric symbolism, and what some theorists now interpret as historical accounts of extraterrestrial contact. At the heart of this interpretive spectrum lies a provocative question: What if these ancient authors were not describing metaphorical angels or theological archetypes, but actual visitors from another world?

The growing popularity of the "ancient astronaut" theory in the 20th and 21st centuries—championed by figures like Erich von Däniken and Zecharia Sitchin—has revived interest in these apocryphal accounts. Supporters of this theory argue that the Watchers were not divine beings in the conventional sense, but rather technologically advanced extraterrestrials whose presence left a permanent mark on human development. Proponents point to descriptions of flying chariots, forbidden knowledge, and cataclysmic judgment as potential evidence of early contact with non-human intelligences. These interpretations, while controversial, find fertile ground in the ambiguous and richly symbolic language of ancient religious literature.

This book explores those possibilities by collecting and modernizing the core texts most frequently cited in ancient astronaut literature, including 1 Enoch, the Book of Giants, Jubilees, and the Testaments of Amram. Alongside these are the lesser-known mystical works of 2 Enoch, 3 Enoch, and the Genesis Apocryphon, which expand the narrative of divine descent, heavenly journeys, and mysterious revelations. Together, these texts offer a panoramic view of ancient cosmologies that defy easy categorization and invite modern re-examination.

Enochian Lore and the Seeds of Speculation

The Book of Enoch, particularly its earliest section—the Book of the Watchers—serves as the foundation for both traditional religious interpretation and modern extraterrestrial speculation. Written in the centuries leading up to the Common Era and once considered canonical by many early Christians, 1 Enoch describes a group of two hundred angels who "descend upon Mount Hermon" to take human wives. These Watchers, led by a figure named Semyaza, not only violate divine law but also teach humanity forbidden arts: metalworking, cosmetics, astrology, enchantments, and other technologies that, to ancient eyes, seemed both miraculous and destructive.

To traditional theologians, these accounts symbolize the dangers of forbidden knowledge, mirroring the transgression of Eden. But to those open to the ancient astronaut theory, the Watchers may represent visitors from the stars—powerful, otherworldly beings who introduced technology and reshaped civilization. The giants born of their union with human women are seen as hybrids—neither fully human nor fully divine, echoing modern fascination with genetic engineering and hybrid species.

The Book of Giants, discovered among the Dead Sea Scrolls, adds a new dimension to this tale by offering fragmented insights into the Nephilim's behavior, dreams, and eventual judgment. These hybrid beings are depicted as possessing immense strength, great size, and terrifying intellects—qualities that resonate with both mythological archetypes and science-fictional alien hybrids. The destruction of the Nephilim and the imprisonment of the Watchers beneath the earth could be interpreted as ancient humanity's collective memory of a divine purge—or a misunderstood attempt to seal away dangerous extraterrestrial beings.

Complementing these texts are the Books of Jubilees and Genesis Apocryphon, which retell the Genesis story with expanded details and deeper cosmological reflection. Jubilees presents a highly structured universe governed by angelic intermediaries, while the Genesis Apocryphon—one of the oldest Dead Sea Scrolls—offers intimate retellings of patriarchal journeys, including visionary dream sequences and supernatural encounters that hint at celestial knowledge.

Cosmic Realms and Extraterrestrial Archetypes

2 Enoch and 3 Enoch, later mystical expansions attributed to the same Enoch of Genesis, transport readers into the heavenly realms. In 2 Enoch, the patriarch ascends through ten levels of heaven, encountering angels, celestial rivers, and divine councils. His transformation from man to angelic being (Metatron in 3 Enoch) adds a new layer of speculative interpretation: might Enoch have been taken aboard a spacecraft, trained, and returned as an emissary of higher intelligence?

In 3 Enoch, the journey continues into more abstract mystical territory, where divine hierarchies and secret names govern spiritual forces. While steeped in Jewish mystical tradition, these descriptions also mirror some modern accounts of alien abduction and enlightenment. The recurring theme of receiving advanced knowledge from beings of light—who dwell beyond visible stars—blurs the line between angel and extraterrestrial.

The Testament of Amram, though preserved only in fragments, reinforces these themes by describing Amram's vision of two beings—one dark, one light—vying for control over human destiny. Their vivid physical descriptions and cosmic roles invite comparisons to ancient dualist cosmologies, such as Zoroastrianism, but also to speculative modern notions of rival extraterrestrial factions influencing earth's history.

These apocalyptic and mystical writings—while varied in tone and theology—share an obsession with descent, revelation, judgment, and hybridization. The boundaries between heaven and earth, spirit and matter, human and divine, are constantly transgressed. It is this transgression that intrigues ancient astronaut theorists, who view such myths as distorted memories of real encounters with superior beings.

Whether or not one accepts the extraterrestrial hypothesis, the texts in this collection remain invaluable as witnesses to humanity's deepest fears and highest aspirations. They speak of our longing to understand our origins, our struggles against corruption, and our hope for redemption—whether that redemption comes from divine grace or distant stars.

In presenting these ancient scriptures through the lens of modern readability and speculative inquiry, this edition invites readers not only to engage with some of the most influential apocalyptic texts of Second Temple Judaism and early Christianity but also to consider the cultural and metaphysical implications of their content. Were the

Watchers fallen angels or fallen astronauts? Were the visions of Enoch divine revelations or coded records of extraterrestrial contact? Were the giants mythological monsters or historical anomalies born of forbidden science?

These questions remain open—not because they defy answers, but because they demand the reader's active participation. You are now invited to explore the stories themselves. Let the Watchers speak. Let the heavens open. Let the flood of mystery begin.

1 Enoch
(Book of the Watchers)

The Books of Enoch, said to be written by the ancient figure Enoch, are important religious texts. Some early Jewish and Christian groups valued them because they describe powerful visions, secrets of the heavens, and messages from God. Even though these books are not included in the Hebrew Bible or most Christian Bibles, they still give insight into what people believed during the time of the Second Temple and early Christianity.

These writings talk about fallen angels, the structure of heaven, and what might happen to humanity in the future. They reveal how people of that era understood the world, using striking images and deep spiritual ideas.

Adding these texts to this collection makes the Apocrypha even richer, allowing readers to explore different ancient religious beliefs. They also raise important questions about God's justice, life's purpose, and how the universe is organized.

Chapter I

Enoch's words of blessing were given to those who are chosen and live rightly. He spoke to those who will be alive during a time of great trouble when all the wicked and godless will be removed.

Enoch, a righteous man, received a vision from God. His eyes were opened, and he saw the Holy One in the heavens. The angels showed him this vision, and he came to understand everything they revealed. But this message was not meant for his own time—it was for a future generation.

He spoke about those who were chosen and delivered this
    message:
The Great and Holy One will leave His dwelling place.
The eternal God will come down to the earth, even to Mount
    Sinai.
He will step out from His camp
And show His great power from the highest heavens.

People everywhere will be filled with fear,
And even the Watchers—mighty beings—will tremble.
Terror will spread across the whole world.
The strongest mountains will shake,
And high hills will be flattened.
They will melt like wax in a blazing fire.

The earth itself will crack apart,
And everything on it will be destroyed.
A great judgment will come upon all people.

But those who live rightly will have peace.
God will protect the ones He has chosen,
And His kindness will surround them.

They will belong to Him completely,
And they will thrive.
They will receive His blessing,
And He will care for each one of them.
His light will shine on them,
And He will fill them with peace.
And look—He will come with tens of thousands of His holy
    ones
To bring justice to all people,

To remove the wicked,

And to judge everyone for their ungodly actions

And the harsh words they have spoken against Him.

Chapter II

Look up at the sky and see how everything moves in order. The stars and other lights follow their paths, rising and setting at the right times. They never change the pattern set for them.

Now look at the earth and notice everything that happens from one end to the other. The land stays firm and steady, never changing. Everything on it continues as it has been, showing the work of God for all to see.

Think about the seasons—summer and winter. See how the earth is covered with water, and how the clouds, dew, and rain settle over the land.

Chapter III

During winter, most trees look dried up and lose all their leaves. But there are fourteen kinds of trees that don't do this. Instead of shedding their leaves, they hold onto them for two to three years until fresh ones grow.

Chapter IV

Watch how the sun sits high in the sky during summer, shining straight down on the land. The heat is so strong that you look for shade to cool off. Even the ground and rocks get so hot that walking on them becomes impossible.

Chapter V

Notice how trees grow fresh green leaves and bear fruit. Pay attention to the world around you and see that the One who lives forever has created everything this way. His works continue in the same cycle, year after year, just as He intended. Everything follows His plan, and nothing changes from what He has commanded.

Look at the seas and rivers too. They follow their course exactly as He directed, never straying from His orders.

But you have not remained faithful. You have not followed the Lord's commands. Instead, you have turned away and spoken with pride, using harsh words against Him. Because of your stubbornness and hardened hearts, you will not find peace.

As a result, you will regret your days, and your years will end in disaster. Your suffering will grow, leading to eternal punishment with no mercy.

During that time, your names will be used as a lasting curse by those who do what is right. People will use your name when they wish harm upon others, and sinners and those without faith will be condemned just as you are. For those who reject God, only a curse remains.

But those who are righteous will be filled with joy. Their sins will be forgiven, and they will receive mercy, peace, and patience. Salvation will come to them, bringing light and hope.

For sinners, there will be no salvation—only a curse. But for the chosen ones, there will be light, happiness, and peace. They will inherit the earth.

Wisdom will be given to them, and they will live without sin. They will not fall into arrogance or wrongdoing, and those who are wise will remain humble.

They will not turn away from what is right, nor will they sin again. They will not die as a result of judgment or anger. Instead, they will live full and peaceful lives, filled with joy. Their happiness will continue forever, and they will enjoy true peace for all their days.

Chapter VI

As people on Earth grew in number, they had daughters who were beautiful. The angels, who came from heaven, saw them and wanted to be with them. They said to each other, "Let's choose wives from among these women and have children with them."

Their leader, Semjâzâ, was unsure and said, "I'm afraid you won't all go through with this plan, and I'll be the only one punished for committing such a great sin." But the others reassured him, saying, "Let's all make a promise together and swear an oath. We will not abandon this plan but will see it through."

So they all swore an oath and made a binding agreement to carry it out. There were two hundred of them in total, and during the time of Jared, they came down to the top of Mount Hermon. They named it Mount Hermon because it was the place where they made their oath and sealed it with a curse.

These were their leaders: Semjâzâ, their chief, along with Arâkîba, Râmêêl, Kôkabîêl, Tâmîêl, Râmîêl, Dânêl, Êzêqêêl, Barâqîjâl, Asâêl, Armârôs, Batârêl, Anânêl, Zaqîêl, Samsâpêêl, Satarêl, Tûrêl, Jômjâêl, and Sariêl. Each of them led a group of ten.

Chapter VII

The others also took wives for themselves, each choosing one. They lived with them and corrupted themselves. They taught the women magical spells, charms, and how to use roots and different plants.

The women became pregnant and gave birth to gigantic children, growing to an unbelievable height. These giants ate everything that people had worked hard to produce. When there was no longer enough food, they turned on the humans and started to eat them instead.

The giants also committed terrible acts against birds, animals, reptiles, and fish. They even began to eat each other's flesh and drink blood. Because of their wickedness, the earth cried out against them.

Chapter VIII

Azâzêl taught people how to make weapons like swords, knives, shields, and armor. He showed them how to work with metals found in the earth. He also introduced jewelry-making, the use of antimony for makeup, and ways to make the eyes look more attractive. He revealed the secrets of precious stones and different colorful dyes.

Because of this, people became more wicked. They fell into immorality, were deceived, and grew more corrupt in their ways. Semjâzâ taught them spells and how to use plant roots for magic. Armârôs showed them how to break enchantments. Barâqîjâl taught astrology, while Kôkabêl revealed the mysteries of the stars. Ezêqêêl explained how to read cloud patterns, Araqiêl taught the signs of the earth, Shamsiêl revealed the secrets of the sun, and Sariêl explained how the moon moves.

As people suffered and died, their cries of pain reached up to heaven.

Chapter IX

Michael, Uriel, Raphael, and Gabriel looked down from heaven and saw that the earth was filled with violence and wrongdoing. They said to each other, "The earth, which was meant to be a peaceful place, is now crying out because of the suffering, and its cries have reached the gates of heaven.

The souls of people are begging us, the holy ones in heaven, to bring their case before the Most High."

Then they spoke to the Lord, saying, "Mighty God, King of kings, ruler over all things, Your throne has stood for all time, and Your name is holy, glorious, and blessed forever.

You created everything, and You have power over all things. Nothing is hidden from You—everything is clear before Your eyes.

You see what Azâzêl has done. He has taught people evil ways and revealed secrets that were meant to stay in heaven—things humans were never supposed to know.

And Semjâzâ, whom You put in charge of his followers, has come down to the earth and taken human women as his own. He and his companions have sinned with them and taught them sinful ways.

Now, their children, the giants, have brought destruction, and the earth is filled with violence and corruption because of them.

The souls of those who have died are crying out for justice, and their voices have reached heaven. Their sorrow will not stop because of all the wickedness happening on earth.

Lord, You knew all of this before it happened. You see what is going on, yet You have not told us what we should do about it."

Chapter X

Then the Most High, the Holy and Great One, spoke and sent Uriel to Lamech's son. He told him, "Go to Noah and warn him in My name. Tell him to hide himself and reveal to him what is about to happen. A great flood will soon cover the earth and destroy everything. Tell him how to survive so that he and his descendants can continue for all generations."

Then the Lord spoke to Raphael, saying, "Capture Azâzêl, tie him up, and throw him into the darkness. Dig a deep pit in the desert of Dûdâêl and cast him into it. Cover him with sharp, jagged rocks and block out all light so that he never sees it again. He will remain there forever. On the day of the final judgment, he will be thrown into the fire. Heal the earth from the corruption caused by the fallen angels so that the plague they brought may end. This way, humanity will not be destroyed because of the forbidden knowledge the Watchers have shared. The whole earth has been ruined by the things Azâzêl has taught, so place all the blame for sin on him."

Then the Lord said to Gabriel, "Go after the evil ones and those born from forbidden unions. Destroy the children of the fallen angels and make them fight each other until they wipe themselves out. They will not live long lives. If their fathers plead for them, do not listen, for they believe they will live forever and expect to reach five hundred years."

The Lord then spoke to Michael, saying, "Go and capture Semjâzâ and his followers, who have taken human women and made themselves unclean with them."

Once their sons have destroyed one another, and they have watched their loved ones perish, bind them for seventy generations deep within the earth. Keep them there until the final judgment, when they will be condemned forever. At that time, they will be cast into the fiery abyss, suffering in a prison where they will be locked away for eternity. Those who are sentenced to destruction will remain with them until the end of all generations.

Wipe out the spirits of the wicked and the children of the fallen angels, for they have harmed humanity. Remove all sin from the earth and put an end to evil. Let righteousness and truth take root and grow. It will bring blessings, and goodness will remain forever in joy and peace.

Then the righteous will be saved. They will live long lives, having thousands of children, and will enjoy both their youth and old age in peace.

After that, the whole earth will be restored to goodness. It will be covered with trees and filled with blessings. Every kind of tree will grow, and vineyards will be planted, producing an abundance of grapes. Crops will yield a thousand times more than before, and olive trees will produce ten times more oil.

Cleanse the world of all oppression, wickedness, and sin. Remove every trace of evil, wiping it completely from the earth.

All people will become righteous, and every nation will give Me honor and praise. They will worship Me together. The earth will be purified from all corruption, sin, and suffering, and I will never again bring such destruction upon it. From generation to generation, the world will remain in peace for all eternity.

Chapter XI

In those days, I will open the great storehouses of blessings in heaven and pour them down onto the earth. These blessings will flow generously, rewarding the hard work of people and enriching their lives. They will not only nourish the land but also bring renewal and prosperity to everyone, filling all of creation with abundance.

Truth and peace will come together in perfect harmony, lasting through every generation. They will be the foundation of life, ensuring that goodness and balance remain forever. This lasting bond between truth and peace will create a world where righteousness thrives, guiding and giving hope to humanity for all time.

Chapter XII

I, Enoch, was giving blessings and praise to the Lord of majesty, the King of all time. As I was doing this, the Watchers called out to me. They spoke to me as Enoch the scribe and said, "Enoch, writer of righteousness, go and deliver a message to the Watchers of heaven—those who left their high and holy home. They have made themselves unclean by taking human wives and acting like the people of the earth.

Tell them, 'You have brought great destruction upon the world. Because of this, you will never find peace, and your sins will never be forgiven. Since you take joy in your children, you will have to watch them die and be destroyed. You will grieve for them and cry out forever, but know this—you will never receive mercy or peace.'"

Chapter XIII

Enoch went to Azâzêl and said, "You will never have peace. A severe judgment has been given against you, and you will be chained. You will not be shown mercy or have your requests granted because of the evil you have taught and the sinful acts you have led people to commit."

Then I went to speak to all of them together, and they were overcome with fear. They trembled in terror and begged me to write a petition for them, hoping they could be forgiven. They wanted me to bring their request before the Lord of heaven.

From that moment, they could no longer speak with Him or even look up toward heaven because they were ashamed of their sins and the punishment they had received. I wrote down their petition, including their prayer about their spirits, their actions, and their plea for forgiveness and a longer life.

I went to sit by the waters of Dan, in the land of Dan, southwest of Mount Hermon. There, I read their petition over and over until I fell asleep.

As I slept, I had a dream and saw visions. I saw punishments being carried out, and a voice called out, telling me to deliver a message to the fallen angels and warn them.

When I woke up, I went to them. They were gathered together, weeping in Abelsjâîl, a place between Lebanon and Sênêsêr. Their faces were filled with shame.

I told them everything I had seen in my dream. Then, I began to speak words of truth and righteousness and rebuked the fallen Watchers for their sins.

Chapter XIV

This book contains words of truth and a warning to the fallen Watchers, as commanded by the Holy Great One in a vision.

While I was asleep, I saw something that I will now share, using the voice and breath given to me by the Great One. He has gifted humanity with speech and understanding so that we can think and communicate. Just as He has given people wisdom, He has also given me the duty to warn the Watchers, the heavenly beings who turned away.

I wrote down your request, but in my vision, I saw that it will not be accepted. Judgment has already been decided, and your plea will never be granted, not now or ever. From this moment on, you will never return to heaven. The decision is final, and you will remain bound to the earth for all time.

You will watch as your beloved sons are destroyed, and there will be no joy left for you. They will die by the sword before your eyes, and your prayers for them will not be heard. Even if you cry out, pray, and repeat every word written in your request, it will not be answered.

Then, in my vision, I saw something incredible: Clouds gathered around me and called me forward, while mist surrounded me. Bright stars and flashes of lightning moved quickly, and powerful winds lifted me, carrying me high into the heavens.

I traveled until I reached a massive crystal wall, surrounded by flames of fire. The sight filled me with fear. I passed through the fire and saw a magnificent house made entirely of crystal. Its walls sparkled like gems, and its foundation was also crystal-clear.

The ceiling of this house looked like the sky filled with stars and lightning, and fiery beings moved between them. The heavens above were as clear as water. A blazing fire surrounded the walls, and the gates of the house glowed with flames.

When I entered, I felt an intense heat, like fire, and at the same time, a deep cold, like ice. There was no comfort inside—only a powerful sense of fear. I trembled and fell on my face.

As I lay there, another vision appeared before me: A second house, even greater and more magnificent than the first, stood open before me. This house was built entirely of fire, and its beauty and size were beyond anything I could describe.

The floor was made of fire, and above it stretched paths of lightning and stars. The ceiling burned with flames. Inside, I saw a high throne that shone like crystal, with wheels as bright as the sun. Around it were visions of heavenly beings.

Beneath the throne, streams of flaming fire flowed so brightly that I could not look at them directly. Seated on the throne was the Great Glory. His robe was brighter than the sun and whiter than any snow. No angel could approach Him because of His overwhelming majesty and brilliance. No living being could look upon His face.

Flames of fire surrounded Him, and a great fire burned before Him. No one could get near Him. Tens of thousands upon thousands stood before Him, yet He needed no advice from anyone. The holiest ones in His presence never left His side, not during the day or the night.

I lay there, trembling, with my face pressed to the ground. Then the Lord Himself spoke, calling my name, "Come here, Enoch, and listen to My words."

One of the holy ones came to me, helped me rise, and led me to the entrance. In deep respect, I bowed my face down before Him.

Chapter XV

Then He answered me, and I heard His voice clearly as He said, "Do not be afraid, Enoch, righteous man and scribe of truth. Come closer and listen carefully to My words.

Go and give this message to the fallen Watchers who sent you to plead for them. It is not humans who should speak on your behalf, but you who should pray for them.

Why did you leave the high and holy heaven? Why did you take

21

human wives and make yourselves unclean with them? You acted like earthly beings and had children with them, creating giants as your sons.

You were once pure and spiritual, living forever, yet you became corrupted by human desires. You gave in to fleshly desires, just like mortal men who are destined to die.

I gave men wives so they could have children and continue life on earth. Their needs would be met, and everything would be provided for them.

But you were different. You were created as spiritual beings, meant to live forever in heaven. That is why I did not give you wives. The heavenly ones were meant to stay in heaven, not take part in human ways.

Now, the children born from your union with human women will be known as evil spirits on earth, for that is where they belong. These spirits come from both men and fallen angels, and because of their corrupted origin, they will forever be known as wicked spirits.

The spirits of heaven will remain in heaven, but the spirits born on earth must stay on earth, as this is their rightful place.

The spirits of these giants will bring suffering, oppression, and destruction. They will fight against people, cause chaos, and bring misery to the world. They will not eat food, but they will always be hungry and thirsty. They will continue to harm others and spread pain. These spirits will rise against humanity and women because they came from them and are tied to their actions."

Chapter XVI

Since the time when the giants were destroyed, the spirits that came from their bodies have continued to bring destruction without being judged. They will keep doing this until the final day—the great day of judgment—when the world as it is now will come to an end. On that day, judgment will be carried out against the fallen Watchers and the wicked, and everything will be set right.

As for the Watchers who sent you to plead for them—those who once lived in heaven—give them this message: "You once lived in heaven, but not all its secrets were revealed to you. Instead, you only learned things that were useless and harmful. Yet, with stubborn hearts, you taught these things to human women. Because of this knowledge, both men and women have done great evil on the earth."

So tell them this: "You will never have peace."

The Book of Giants

Introduction

The mysterious figure of Enoch, briefly mentioned in the Book of Genesis, has fascinated scholars and religious thinkers for centuries. The Bible describes him as a man who "walked with God" and was taken by Him (Genesis 5:24), which has inspired many writings that explore his life and the mysteries surrounding him. One of these texts, The Book of Giants, gives a deeper look into the world before the Great Flood, focusing on events involving powerful beings known as the Watchers.

The journey of The Book of Giants from being almost unknown to gaining scholarly attention shows how ancient stories can survive over time. Pieces of this text were first discovered among the Dead Sea Scrolls in the mid-20th century, specifically in caves 1, 2, 4, and 6 at Qumran. These Aramaic fragments, written before the 2nd century BCE, helped connect the short biblical mentions of Enoch with the more detailed stories found in other ancient writings. The discovery highlighted the text's importance in understanding the religious and cultural world of the Second Temple period.

The Book of Giants tells the story of what happened when the "sons of God" came to Earth. These heavenly beings, called the Watchers, formed forbidden relationships with human women, which led to the birth of giant hybrid children known as the Nephilim. These giants, blessed with incredible strength and size, soon became violent rulers, bringing chaos and destruction to the world. The story focuses on two of these giants, Ohyah and Hahyah, the sons of the leader of the Watchers, Shemihaza. Their dreams, later explained by

Enoch, warn of a coming disaster as punishment for their actions. The book highlights themes of divine justice and the dangers of breaking the natural order.

Many of the ideas in The Book of Giants are similar to stories from other ancient cultures, especially myths about gods or divine beings having children with humans. The presence of names like Gilgamesh, a well-known figure from Mesopotamian legends, suggests that the book combines elements from different traditions, showing a shared cultural history. The text also adds more details to the biblical story of the Great Flood, explaining that it was not just caused by human evil but also by the chaos created by the Watchers and their children.

Although The Book of Giants was written before the 2nd century BCE, it was kept alive through the Manichaean religion. Manichaeism, a gnostic faith that began in the 3rd century CE, adopted and modified the story, making it part of their religious teachings. Fragments of the book have been found in Turfan, Western China, proving that it was shared across many cultures and remained influential for centuries.

Today, The Book of Giants gives scholars and history enthusiasts a better understanding of early Jewish beliefs, ancient religious writings, and the development of ideas about angels and demons. The stories in this book challenge readers to think about the divide between the human and divine worlds—and what happens when those boundaries are crossed. As we explore this translation, we have the opportunity to reflect on these ancient ideas and what they say about morality, justice, and the human experience.

This version of The Book of Giants aims to stay true to its original meaning while making it easy to read and understand. It keeps the richness of the original text while helping readers see the historical and cultural background that shaped its creation.

Book Of Giants -- Reconstructed Texts

A group of fallen angels came down to Earth, bringing both secret knowledge and destruction.

They learned things they were never supposed to know.
Sin spread everywhere.
They became violent and killed many people.
Their children grew into giants.

The angels used Earth's resources for themselves.

They took everything the land produced.
They ruled over the sea, the sky, and all living things.
They ate the fruit, grain, and trees.
They took animals, from wild beasts to tiny creatures, and
    watched everything happening on Earth.
They joined in terrible acts, bringing corruption to humanity.

The two hundred fallen angels began experimenting with
    different creatures, including humans.

They took donkeys, rams, goats, and other animals from the
    land and sky.
They used them in unnatural ways, mixing them in ways that
    were never meant to happen.

This corruption led to chaos, violence, and monsters roaming
    the Earth.

They ruined the world.

They created giants and unnatural creatures.

Their evil spread everywhere.

The Earth was filled with bloodshed, and the giants were never
satisfied.

They destroyed everything and devoured whatever they could.

Monsters attacked whatever stood in their way.

The land was torn apart.

Terrifying creatures appeared and grew stronger.

They didn't understand the full impact of their actions.

Their choices made the Earth worse.

The fallen angels' influence caused great destruction.

In the end, everything would collapse because of them.

Yet, even after all this, they were never satisfied.

The giants began having strange dreams and visions. One of them,
Mahway, the son of the angel Barakel, had a dream that disturbed
him. He saw a tablet placed in water. When it was pulled out, only
three names remained while the rest had disappeared. This seemed to
warn that almost everyone would be wiped out, leaving only a few
survivors—Noah and his family—after a great flood.

A tablet was soaked in water.

The water rose and covered it.

When it was lifted out, most of the names were gone.

Mahway told the other giants about his dream, and they tried to
understand what it meant.

He realized the vision was a warning of disaster.

He admitted his fear to the others and spoke of the spirits of the

dead, who cried out for justice against those who had killed
them.
He saw that they would all die soon, and their time was almost
up.

Ohya, one of the giants, asked Mahway who had given him this
vision.
Mahway said his father, Barakel, had been with him.
Before he could finish, Ohya interrupted, shocked by what he
had heard.
He shouted, "This is unbelievable! If even a woman who cannot
have children gives birth, then something truly impossible is
happening."

Ohya spoke to Hahya, saying that destruction was coming to the
Earth.
When they realized this, they cried before the giants.

Ohya then told Hahya that it was not their fault, but Azaiel's.
He said the giants were the children of fallen angels, and they
would not let their loved ones be abandoned.
He also said they had not been completely defeated, and they
still had strength left.

The giants began to understand that they could not win against
the powers of Heaven. One of the speakers might have been
Gilgamesh.

A giant declared that he was strong, with great power in his
arms.
He had fought against mortals and waged war, but he now

realized he could not defeat his enemies.
They lived in Heaven, in sacred places, and were far stronger
than him.
He had been called a wild beast and a wild man.

Then Ohya spoke and said he had a dream that disturbed him.
It kept him awake and forced him to see a vision.
He now understood something important.
His vision was about a tree being uprooted, except for three
roots.
It carried the same message as a previous dream.

As he watched, three roots remained.
Then something moved them into a garden.

Ohya tried to ignore what the vision meant.
At first, he claimed it only referred to the demon Azazel.
Now, he suggested it was only about the rulers of the Earth.

The vision spoke about the fate of their souls.
Ohya told the giants what Gilgamesh had said.
The leader had cursed the rulers, and the giants were pleased by
his words.
Then he left.

The giants were troubled by more dreams.
Two of them woke up, frightened by what they had seen.
They went to their fellow giants and described their visions.

One of them had dreamed of a garden.
Gardeners were watering trees.
Two hundred trees had large branches growing from their roots.

Then, fire spread and burned the entire garden.

The dreamers went to the giants and told them what they had
     seen.
Someone suggested they should find Enoch to explain the
     meaning of the dreams.

Ohya spoke to the giants and told them about his own dream.
He had seen the Ruler of Heaven come down to Earth.
When he finished, the giants and monsters became terrified.
They called Mahway and asked him to go find Enoch.

Mahway was sent to find Enoch, the wise scribe, to ask him
     about the visions.
He flew through the air like strong winds, moving like an eagle.
He left the land behind and passed through a great desert.

When Enoch saw Mahway, he greeted him.
Mahway told him the giants and monsters were waiting for
     answers.
They wanted to know what their dreams meant.
Enoch sent back a tablet with a warning.
The message was written in his own hand and was addressed to
     Shemihaza and his followers.

He warned them about the things they had done.
Their wives and children had followed their wicked ways.
The land itself was crying out against them.
Because of their actions, destruction was coming.
A great flood would wipe out all life on land and sea.

But there was still a chance to change.

They were told to break free from evil and pray.

In another vision, Enoch saw something that filled him with
fear.
He collapsed to the ground when he heard a voice.
He saw a being who lived among humans but had not learned
from them.

The Book of Jubilees

Introduction

The Book of Jubilees, also known as "The Little Genesis," is an ancient Jewish text that expands on the stories in Genesis and Exodus. It was likely written between the 2nd and 1st centuries BCE and retells biblical history using a unique system of time—dividing events into periods of forty-nine years, called jubilees. This structure provides a detailed timeline from creation up to the moment when God gave the law at Mount Sinai. The book also gives insight into ancient Jewish beliefs, laws, and traditions.

What makes Jubilees unique is its strong focus on the idea that God's laws came directly from Him, the importance of the Sabbath, and the role of angels in delivering God's messages. The text reflects the beliefs and customs of a particular Jewish group, possibly connected to the Essenes. While it is not included in the Hebrew Bible, it is considered a sacred text in the Ethiopian Orthodox Church and is valued for its deep religious and historical meaning.

This important book acts as a link between the Bible and later writings about the end times, giving us a better understanding of biblical stories and the culture and religious practices of that time.

Chapter I.

In the first year after the Israelites left Egypt, on the sixteenth day of the third month, God spoke to Moses, saying, "Come up to Me on the mountain, and I will give you two stone tablets with My laws and commandments. You will teach them to the people."

Moses went up Mount Sinai, and God's glory covered the mountain in a cloud for six days. On the seventh day, God called to Moses from the cloud. His presence on the mountaintop looked like a blazing fire. Moses stayed on the mountain for forty days and forty nights, and during this time, God showed him past and future events, organizing all the laws and teachings.

God said, "Pay close attention to everything I am telling you and write it down in a book. In the future, people will realize that even when they sin and break My covenant, I have not abandoned them. When these events take place, they will understand that My judgments are right and fair. They will see that I have always been with them."

Write down everything I tell you today because I already know how stubborn and rebellious they will be. Even before I bring them into the land I promised to their ancestors—Abraham, Isaac, and Jacob—they will turn away from Me. They will enjoy all its blessings, eat until they are full, and then follow false gods that cannot save them when trouble comes. This will stand as a witness against them. They will forget My commandments and follow the sinful ways of the nations around them, worshiping idols and practicing evil. These false gods will become a trap and a burden to them.

Many will die or be taken captive by their enemies because they rejected My laws. They will stop celebrating My holy days, break My Sabbaths, and abandon the sacred place I gave them. Instead, they will build altars on high places, worship idols, and even sacrifice their children to demons. They will do terrible things because of the evil in their hearts.

I will send messengers to warn them, but they will refuse to listen. They will kill these messengers, persecute those who follow My law, and twist My words to justify their wrongdoing. Because of this, I will turn away from them and allow foreign nations to capture them and destroy their land. They will be scattered among different nations, and while in exile, they will forget My laws and commandments. They will lose understanding of My holy days and drift even further from Me.

But one day, they will return to Me with all their heart, soul, and strength. When they truly seek Me, they will find Me. I will bring them back from the nations where they were scattered and give them peace and righteousness. I will fill them with goodness and bless them instead of cursing them. They will no longer be the oppressed but the leaders.

I will place My sanctuary among them and live with them. I will be their God, and they will be My people, walking in truth and righteousness. I will never leave them because I am their Lord and God.

Moses fell on his face and prayed, saying, "O Lord, do not abandon Your people, Your chosen ones. Do not let them fall into the hands of their enemies, who will lead them further into sin. Show them mercy, Lord. Create a pure spirit within them so they do not continue down the path of evil and perish before You. They are Your people, whom You saved with great power from Egypt. Give them clean hearts and holy spirits so they do not fall into sin again."

The Lord answered Moses, "I know how stubborn they are. They will not fully obey Me until they admit their sins and the sins of their ancestors. But when they return to Me with all their heart and soul, I will change them. I will give them new hearts, and their children will follow Me as well. I will fill them with My holy spirit and purify them, so they will never turn away from Me again. They will obey My laws, and I will be their Father, and they will be My children. Everyone in heaven and on earth will know that they are My people, and I am their God. I love them with an everlasting love.

Write down everything I am telling you—past, present, and future. These words will stand for all generations, guiding them until I come to live among them forever."

Then God said to the angel of His presence, "Write everything down for Moses, from the beginning of creation until the time My sanctuary will be built forever among them." The Lord will reveal Himself to all, and everyone will know that He is the God of Israel, the Father of Jacob's descendants, and the eternal King who reigns from Mount Zion. Jerusalem and Zion will be holy forever.

The angel of the presence, who guided Israel through the wilderness, brought the tablets containing the history of the world—from the creation of time to the final renewal of heaven and earth. All creation will be restored as it was meant to be, and the Lord's sanctuary will be established in Jerusalem on Mount Zion. The stars and heavenly lights will be renewed, bringing healing, peace, and blessings to God's chosen people forever.

Chapter II.

Then the angel, following God's command, spoke to Moses and said, "Write down the full story of creation. Record how, in six days, God made everything and brought it to life. On the seventh day, He rested and made it a special, holy day for all time.

On the first day, God created the sky, the earth, and the waters. He also made angels—some to be in His presence, some to bring holiness, and others to control fire, wind, clouds, snow, hail, and frost. He created angels for thunder, lightning, and the changing seasons. He also made spirits for all His creatures, both in heaven and on earth. He formed the deep waters, darkness, evening, night, light, dawn, and daytime. Everything was made with His wisdom. We saw His creation and praised Him. Seven great things were made on the first day.

On the second day, God made the sky and placed it between the waters. Some waters rose above the sky, while others remained below, covering the earth. This was the only thing He created on the second day.

On the third day, God commanded the waters under the sky to come together so that dry land would appear. The waters obeyed, forming seas, rivers, and lakes. On this day, He also created dew, seeds, and plants. He made fruit trees, forests, and the Garden of Eden, filled with all kinds of plants. Four important things were made on the third day.

On the fourth day, God made the sun, moon, and stars. He placed them in the sky to shine on the earth, to separate day from night, and to mark time. These lights were also signs for the days, the Sabbath, the months, festivals, years, and special cycles of time. Three great things were made on the fourth day.

On the fifth day, God created the great sea creatures that live in the deep waters. These were the first living things He made. He also created fish and all creatures that live in water, as well as all kinds of birds. When the sun rose, it shone on these creatures and blessed them, along with all the plants and trees that grow on the earth. Three kinds of living beings were made on the fifth day.

On the sixth day, God made land animals, including livestock and creatures that move on the ground. After that, He created humans, making both man and woman. He gave them control over the earth, the seas, the birds, the animals, and all living things. They were put in charge of everything on earth. Four types of creation were made on the sixth day, bringing the total to twenty-two.

On this day, God finished all His work—the heavens, the earth, the seas, and everything in them. He established a special sign: the

Sabbath. He commanded that people should work for six days and rest on the seventh.

God also told His angels to observe the Sabbath with Him, both in heaven and on earth. Then He said, "I will choose a special people from all the nations, and they will keep the Sabbath. I will make them My people and bless them, just as I have blessed and set apart this day for Myself. They will belong to Me, and I will be their God.

From everything I have seen, I have chosen Jacob's descendants as My firstborn son. I have set them apart forever and will teach them to honor the Sabbath, so they may rest and keep it holy."

That is why the Sabbath is a sign—a day to celebrate with food, drink, and praise to the Creator. Just as God chose a special people, they will keep the Sabbath and celebrate with us.

His commandments were given as a way to praise Him forever.

From Adam to Jacob, there were twenty-two generations, just like there were twenty-two kinds of work completed before the seventh day. The Sabbath was blessed along with the days before it, making it a time of holiness and rest.

To Jacob and his descendants, God gave the promise that they would be a holy and blessed people. This was part of His first law and covenant, just as He blessed the Sabbath.

In six days, God created the heavens, the earth, and everything in them. On the seventh day, He made it holy. He commanded that anyone who works on this day must be punished, and anyone who disrespects it will suffer.

Teach the Israelites to keep this day holy and rest from all work, for it is the most sacred of all days. Whoever disrespects it will be punished, and whoever works on it will face consequences forever. This law was given so that the Israelites would always observe the

Sabbath and never lose their inheritance. It is a holy and blessed day.

Those who honor it and rest will also be holy and blessed, just as we are.

Tell the Israelites to always keep the Sabbath. Let them know that on this day, they should not do unnecessary work, seek their own pleasure, prepare food or drink, fetch water, or carry heavy loads through their gates. All of this must be done on the sixth day.

They should also not move things between houses on the Sabbath. This day is even more sacred than a jubilee. We in heaven have been observing the Sabbath long before it was given to humans.

The Creator blessed this day, but He did not require every nation to follow it. He set Israel apart to keep this law. Only they were chosen to eat, drink, and celebrate the Sabbath on earth.

The Creator made this day special, setting it apart as the holiest and most honored of all days.

This law was given to the Israelites as a lasting command for all generations.

Chapter III.

During the second week, over six days, we brought all kinds of animals to Adam. On the first day, he saw wild animals, on the second, livestock, on the third, birds, on the fourth, land creatures, and on the fifth, sea creatures.

Adam gave each one a name, and whatever he called them became their name. Over those days, he saw every kind of animal, both male and female, but he remained alone—there was no companion for him.

Then the Lord said to us, "It is not good for man to be alone. Let us make a helper for him."

So, God put Adam into a deep sleep. While he slept, God took one of his ribs and used it to create a woman. Then He closed the place where the rib had been removed.

When Adam woke up on the sixth day, God brought the woman to him. Adam immediately recognized her and said, "She is part of me—bone of my bones, flesh of my flesh. She will be called 'woman' because she was taken from man."

This is why a man leaves his parents and joins his wife, and the two become one.

Adam was created in the first week, and in the second week, the woman—made from his rib—was brought to him. God showed her to him, and because of this, a command was given: if a woman gives birth to a boy, she will be unclean for seven days, but if she gives birth to a girl, she will be unclean for fourteen days.

After Adam had lived in the land where he was created for forty days, we took him to the Garden of Eden so he could take care of it. His wife was brought into the Garden on the eightieth day, and from then on, they lived there together.

For this reason, a law was written on the heavenly tablets: "When a woman has a son, she will be unclean for seven days, just like the first week. Then, for thirty-three more days, she must stay away from anything holy and cannot enter the sacred place until her purification is complete."

If she has a daughter, she will be unclean for two weeks, just like the first two weeks, and will need sixty-six more days to complete her purification, making a total of eighty days.

Once these eighty days are over, she may enter the holy place again, because the Garden of Eden is holier than the rest of the earth, and every tree in it is sacred.

This is why the law was given regarding childbirth: a woman must not touch anything holy or enter the sacred place until her purification is complete.

This law was recorded for Israel so they would follow it for all generations.

During the first week of the first jubilee, Adam and his wife lived in the Garden of Eden for seven years. They took care of it, following the instructions they were given. Adam worked hard, and though he was naked, he felt no shame. He protected the garden from birds, animals, and livestock. He gathered fruit and saved some for himself and his wife.

After exactly seven years, in the second month, on the seventeenth day, the serpent approached the woman. It asked, "Did God really say you cannot eat from any tree in the garden?"

The woman replied, "We can eat from any tree except the one in the middle. God said, 'Do not eat from it or even touch it, or you will die.'"

The serpent said, "You won't die. God knows that if you eat it, your eyes will be opened, and you will be like gods, knowing good and evil."

The woman looked at the tree and saw that its fruit looked good to eat, was beautiful, and seemed to bring wisdom. She took some, ate it, and gave some to Adam. He ate as well.

Immediately, their eyes were opened, and they realized they were naked. They sewed fig leaves together to cover themselves.

God cursed the serpent, giving it eternal punishment. Then He turned to the woman and said, "Because you listened to the serpent and ate the fruit, I will greatly increase your pain in childbirth. You will suffer when you have children. You will long for your husband,

and he will rule over you."

To Adam, He said, "Because you listened to your wife and ate from the tree I told you not to eat from, the ground is now cursed. You will have to work hard to get food from it for the rest of your life. It will produce thorns and weeds, and you will eat the plants of the field. You will sweat as you work for your food until the day you return to the ground. You were made from dust, and you will return to dust."

Then God made clothes from animal skins for Adam and his wife and dressed them. After that, He sent them out of the Garden of Eden.

On the day Adam left the garden, he offered a sacrifice at sunrise, burning fragrant spices like frankincense, galbanum, and stacte to seek forgiveness for his shame.

That same day, all animals, birds, and creatures that moved on the earth became silent. Before this, they had all spoken the same language. Then God sent all the creatures out of the Garden of Eden, separating them into their proper places. Of all the living things, only Adam was given clothing to cover his nakedness.

This is why it is written on the heavenly tablets that all who know the law's judgment must cover themselves and not expose their bodies as the other nations do.

On the new moon of the fourth month, Adam and his wife left the Garden of Eden and settled in the land of Elda, the place where they were created. Adam named his wife Eve.

During the first jubilee, they had no children. Later, Adam was with his wife, and he worked the land just as he had been taught in the Garden of Eden.

Chapter IV.

During the third week of the second jubilee, Eve gave birth to Cain. In the fourth week, she had Abel, and in the fifth week, she gave birth to their daughter, Âwân.

In the first year of the third jubilee, Cain killed Abel because God accepted Abel's offering but rejected his. Cain attacked Abel in a field, spilling his blood, which cried out to heaven for justice.

God confronted Cain about his crime, and as a result, Cain was cursed and became a wanderer. He had to live with the guilt of his brother's death. This is why it is written on the heavenly tablets: "Cursed is anyone who kills their neighbor in secret, and all who hear of it must say, 'So be it.' Those who stay silent share in the guilt."

This is why we confess all our sins before the Lord—whether they happen in heaven, on earth, in the open, or in secret—so that nothing remains hidden. Adam and Eve mourned Abel for many years. But in the fourth year of the fifth week, their sadness turned to joy when Adam was with his wife again, and she gave birth to Seth. Adam said, "God has given us another child to take Abel's place."

In the sixth week, Eve had a daughter named Azûrâ. Cain married his sister Âwân, and by the end of the fourth jubilee, she gave birth to their son, Enoch. In the first year of the first week of the fifth jubilee, people began building houses on the earth. Cain built a city and named it after his son, Enoch. Adam and Eve had nine more sons.

During the fifth week of the fifth jubilee, Seth married his sister Azûrâ, and in the fourth year of the sixth week, they had a son named Enos. Enos was the first to call on the name of the Lord.

In the third week of the seventh jubilee, Enos married his sister Nôâm, and in the third year of the fifth week, they had a son named

Kenan. At the end of the eighth jubilee, Kenan married his sister Mûalêlêth, and in the third year of the first week of the ninth jubilee, they had a son named Mahalalel.

During the second week of the tenth jubilee, Mahalalel married Dinah, the daughter of Barakiel, who was his cousin. In the sixth year of the third week, they had a son named Jared.

During Jared's time, a group of angels called the Watchers came down to teach people about justice and judgment. In the eleventh jubilee, Jared married Baraka, the daughter of Râsûjâl, another relative. In the fifth week of that jubilee, she gave birth to Enoch.

Enoch was the first to learn writing, wisdom, and knowledge. He studied the signs in the sky, helping people understand time and seasons. He recorded the weeks, years, and Sabbaths exactly as they were revealed to him. He also received visions of past and future events, writing them down for future generations.

In the twelfth jubilee, during the seventh week, Enoch married Edna, the daughter of Danel, his cousin. In the sixth year of that week, they had a son named Methuselah.

Enoch spent six jubilees with the angels of God, learning about heaven and earth. He wrote everything down and warned against the Watchers, who had taken human wives. Because of his righteousness, God took Enoch away from the world and placed him in the Garden of Eden, where he recorded His judgments.

Later, God sent a flood to cleanse the earth because of human wickedness. Enoch's life stood as a warning to people. He also burned incense on a mountain, offering sweet-smelling spices to God. The Lord established four sacred places on earth: the Garden of Eden, the Mount of the East, Mount Sinai, and Mount Zion. These places would one day be made holy again to cleanse the world from its corruption.

In the fourteenth jubilee, Methuselah married Edna, the daughter of Azrial, his cousin. During the third week, in the first year of that week, she gave birth to Lamech.

In the fifteenth jubilee, Lamech married Betenos, the daughter of Baraki'il, his cousin. They had a son named Noah. Lamech said, "This child will bring us relief from the hardship caused by the cursed ground."

At the end of the nineteenth jubilee, during the seventh week of the sixth year, Adam died. His sons buried him in the land where he had been created. He lived 930 years but did not reach 1,000. According to the heavenly record, 1,000 years is like one day, so the prophecy about the tree of knowledge was fulfilled: "On the day you eat from it, you will die." In God's time, Adam died within that same "day."

That same year, Cain was killed when his house collapsed on him. This fulfilled what was written on the heavenly tablets: "Whoever kills with a weapon will be killed by the same." Since Cain killed Abel with a stone, he also died under stones as a just punishment.

In the twenty-fifth jubilee, Noah married Emzârâ, the daughter of Râkê'êl, his cousin. She gave birth to three sons: Shem in the third year, Ham in the fifth year, and Japheth in the first year of the sixth week.

Chapter V.

When the number of people on earth increased and they had daughters, the angels of God noticed how beautiful they were. During a certain year in a jubilee, they chose wives for themselves from among them, taking whoever they wanted. Their children grew into giants, and soon, lawlessness spread across the land. Corruption filled the world as humans, animals, and birds became violent, attacking and even eating each other. People's thoughts became completely evil, and wrongdoing took over everywhere.

God saw how wicked the world had become. All living creatures had strayed from His ways, and the earth was filled with sin. Because of this, He decided to destroy mankind and every living thing He had created. But Noah stood apart—he was righteous, and God looked upon him with favor.

God's anger also turned toward the angels He had sent to earth. He stripped them of their power and commanded that they be chained deep underground, cut off from the rest of creation. Their giant offspring were also condemned to die. God declared that they would be wiped out by the sword and removed from the earth. He said, "My Spirit will not remain with humans forever, for they are mortal. Their lifespan will now be 120 years." Then, He caused violence to spread among the people, making them turn against each other until they were completely destroyed.

The fallen angels, forced to watch their children die, were imprisoned in the depths of the earth. There they will remain until the final judgment when God will punish all those who corrupted the world. No one escaped judgment. The wicked were completely removed from the earth. Afterward, God gave a new, pure nature to all living creatures so they would not turn back to evil.

The heavenly tablets record the laws for all creation. It is written that anyone who strays from their rightful path will be judged. Every action—whether done in heaven, on earth, in darkness, in light, in the depths, or even in Sheol—is seen by God. He is a just judge who does not show favoritism or accept bribes. If He has decided on a judgment, He will carry it out completely.

However, God also promised that if Israel repents and turns back to Him, He will forgive their sins. Once a year, He will grant mercy to those who admit their guilt and change their ways. But those who became corrupt before the flood were given no mercy. Only Noah was accepted by God, and his righteousness saved not only himself but also his sons. He followed all of God's instructions exactly as he was told.

God decided to destroy everything on earth—humans, animals, and birds. He commanded Noah to build an ark to survive the flood. Noah obeyed, constructing the ark just as God had instructed. This happened in the twenty-seventh jubilee, during the fifth week, in the fifth year, on the new moon of the first month. In the sixth year, during the second month, Noah and the animals entered the ark. On the seventeenth evening, the Lord shut the door from the outside.

Then, God opened the seven floodgates of heaven and the seven deep springs of the earth. For forty days and nights, heavy rain poured down, and water gushed up from the ground. The floodwaters rose until they covered even the tallest mountains by fifteen cubits. The ark floated on the surface of the waters. For five months—150 days—the flood remained over the earth. Finally, the ark came to rest on Mount Lubar, one of the mountains in the Ararat range.

In the fourth month, the deep springs of the earth were sealed shut, and the floodgates of heaven were closed. By the seventh month, the waters began to drain into the earth's depths. In the tenth month, the mountain peaks became visible again. By the first month of the new year, the land started to dry. On the seventeenth day of the second month, the ground was completely dry. Then, on the twenty-seventh day, Noah opened the ark, letting all the animals, birds, and creatures go free to return to the land.

Chapter VI.

On the first day of the third month, Noah left the ark and built an altar on the mountain. He made a sacrifice to cleanse the earth, taking a young goat and using its blood to remove the guilt of the land. Everything that once lived on the earth had been wiped out, except for those saved in the ark with Noah. He placed the fat of the sacrifice on the altar and also offered an ox, a goat, a sheep, young goats, salt, a turtledove, and a young pigeon. He poured oil over them, sprinkled wine, and burned frankincense, creating a pleasing aroma that rose to God.

The Lord smelled the offering and made a promise to Noah, vowing never to destroy the earth by a flood again. He said that planting and harvest, cold and heat, summer and winter, day and night would continue as they were meant to, without interruption. Then He told Noah, "Have many children and fill the earth. I have placed all animals on land, in the sea, and in the sky under your authority. Just as I gave you plants to eat, I now give you everything. But do not eat meat that still has blood in it, because life is in the blood. Anyone who takes another person's life will be held accountable, for humans were made in God's image. Be fruitful and spread across the earth."

Noah and his sons made a vow before God that they would never eat blood. This became a lasting agreement for all future generations. God commanded that Israel remember this covenant, just as He later instructed Moses on the mountain. They were to sprinkle blood as part of their worship and never eat blood from any animal, bird, or livestock. This law was given as a permanent reminder so that Israel would always keep this commandment. Anyone who ate blood would be cut off from the land, and their descendants would be forgotten before God.

To seal His promise, God gave Noah a sign—He placed a rainbow in the clouds as a symbol of His eternal covenant that He would never again flood the earth to destroy it. It was also recorded in the heavenly tablets that a special festival, the Feast of Weeks, would be celebrated every year in this month to renew the covenant. This festival had been observed in heaven from the creation of the world until Noah's time, lasting 26 jubilees and five weeks of years. Noah and his sons continued to celebrate it for seven jubilees and one week of years until Noah died. After that, people forgot about it until Abraham revived it. His son Isaac, then Jacob, and their descendants observed it until Moses renewed it on the mountain.

God commanded the people of Israel to keep this festival in every generation, celebrating it as a reminder of the covenant. It became known as both the Feast of Weeks and the Feast of Firstfruits, a special day written into the law. It was established that this festival would be celebrated on a specific day each year, and Moses was given instructions on how it should be observed, ensuring that the Israelites would keep it faithfully for generations.

The new moons of the first, fourth, seventh, and tenth months were also set as days of remembrance, marking the changing seasons and the divisions of the year. Noah established these days as lasting feasts. On the new moon of the first month, he was told to build the ark, and on that same day, the earth became dry after the flood. On the new moon of the fourth month, the deep springs of the earth were sealed shut. On the new moon of the seventh month, the earth's abysses opened, allowing the floodwaters to drain. On the new moon of the tenth month, the mountain peaks became visible, and Noah rejoiced. These days were recorded in the heavenly tablets and were meant to be remembered forever.

The year was divided into four seasons, each lasting 13 weeks, making a total of 52 weeks. This cycle was written and established in the heavenly tablets and was not to be changed. The Israelites were commanded to follow a 364-day year, so the feasts and seasons would remain in their correct order. Any changes to this system would throw the years out of sync, leading to confusion about the feasts, new moons, Sabbaths, and the seasons.

God warned Moses that after his death, the Israelites would eventually stray from this system. They would abandon the 364-day calendar, causing their festivals, holy days, and Sabbaths to fall out of order. Some would begin following the lunar calendar, leading to mistakes in their observances and making holy days unclean. They would mix the sacred with the ordinary, forget the commandments, and even start eating blood and all kinds of meat without following the laws.

This message was given as a warning, so the Israelites would not be led astray and so they would understand the serious consequences of ignoring the appointed times and the laws of the covenant.

Chapter VII.

In the seventh week of the first year of this special time, Noah planted vineyards on the mountain where the ark had landed—Mount Lubar, one of the Ararat Mountains. It took four years for the vines to grow and produce grapes. In the seventh month of that year, Noah gathered the grapes.

Noah made wine from the grapes and stored it in a container. He kept it until the fifth year, and on the first day of the first month, he held a big feast. As part of the celebration, he made a burnt offering to God, sacrificing a young ox, a ram, seven one-year-old sheep, and a young goat to ask for forgiveness for himself and his sons.

Noah started with the goat, using some of its blood and meat on the altar. He placed all the fat on the altar as part of the burnt offering.

Then, he did the same with the ox, the ram, and the sheep, placing their meat and fat on the altar and mixing them with oil. He poured wine over the fire and burned incense. The smell rose up and pleased God.

Noah and his sons enjoyed the feast, drinking the wine with joy. As the evening came, Noah went to his tent, fell asleep, and became drunk. While he was asleep, he accidentally uncovered himself and lay naked.

His son Ham saw him like this and went to tell his brothers. But Shem and Japheth took a garment, placed it on their shoulders, and walked backward into the tent to cover their father without looking at him.

When Noah woke up and realized what had happened, he cursed Ham's son, Canaan, saying, "Canaan will be a servant to his brothers." Then he blessed Shem, saying, "Praise be to the Lord, the God of Shem. May Canaan serve him." He also prayed for Japheth, saying, "May God bless Japheth and let him live among Shem's people. And may Canaan be his servant too."

Ham was upset when he heard the curse on his son. He left Noah and went with his sons—Cush, Mizraim, Put, and Canaan—to build a city, naming it after his wife, Ne'elatama'uk.

Japheth, feeling jealous, built his own city and named it after his wife, 'Adataneses.

Shem, however, stayed close to Noah and built a city near him on the mountain, calling it Sedeqetelebab, after his wife.

This is how three cities were built near Mount Lubar: Sedeqetelebab in the east, Na'eltama'uk in the south, and 'Adataneses in the west.

The sons of Shem were Elam, Asshur, Arpachshad (who was

born two years after the flood), Lud, and Aram. Japheth's sons were Gomer, Magog, Madai, Javan, Tubal, Meshech, and Tiras. These were the children and grandchildren of Noah.

In the twenty-eighth jubilee, Noah began teaching his grandsons the laws and commands he had received. He told his sons to live good and honest lives, to dress modestly, respect their Creator, honor their parents, love their neighbors, and avoid evil and impurity.

He warned them that the flood had come because of three great sins: wickedness, impurity, and the wrongdoing of the Watchers, who had taken wives from among human women.

These sins led to the birth of the Nephilim, who became violent and corrupt. The Giants killed the Nephilim, the Nephilim killed the Eljo, and the Eljo turned against humans. The world became filled with wickedness, violence, and bloodshed.

People didn't just harm one another; they also hurt animals, birds, and all living creatures. Blood covered the earth, and people's hearts became filled with evil thoughts and wicked plans.

Because of this, God wiped everything off the face of the earth. Every living thing was destroyed because of the violence and corruption.

Noah then spoke to his sons, saying, "I see what you are doing, and it worries me. You are not following the right path. You are becoming jealous of one another, arguing, and growing apart. I fear that after I die, you will start shedding blood, and because of this, you too will be removed from the earth.

Anyone who kills another person or drinks animal blood will be destroyed. Their family line will end, and they will have no descendants. Those who do such things will go to Sheol, a place of punishment and darkness, and they will be taken from the earth.

The killing of people and animals must stop. Any blood spilled must be covered because I have been commanded to warn you, your children, and all creatures.

Do not let your hearts be stained by the flesh of animals, for blood is life, and God will hold everyone accountable for the blood they shed. The earth cannot be cleansed of spilled blood unless the one who caused it also sheds their own blood. Only then will the land be pure.

Now, my children, listen to me. Live righteously and fairly, so that you will always walk in the right way. If you do good, your actions will rise before God, who saved me from the flood.

Go and build cities, plant trees, and grow crops. When you plant fruit trees, do not pick the fruit for the first three years. In the fourth year, the fruit will be holy, and the first portion should be given to God, the Creator of heaven and earth.

The remaining fruit will be for those who serve God. In the fifth year, you may harvest the fruit freely, and everything you plant will grow well.

This rule was first given by Enoch, the ancestor of your forefathers. Methuselah passed it down to his son Lamech, and Lamech taught me, just as their ancestors had taught them.

Now, I am passing this rule to you, just as Enoch gave it to his son long ago. He carefully taught his children and grandchildren, ensuring they followed it throughout their lives."

Chapter VIII.

In the first week of the twenty-ninth jubilee, at the very beginning, Arpachshad married Rasu'eja, the daughter of Susan, who was Elam's daughter. Three years later, they had a son named Kainam. As

Kainam grew older, his father taught him how to write. One day, Kainam went out to look for a place to build a city. While exploring, he found an old inscription carved into a rock by earlier generations. He read what was written, copied it down, and unknowingly committed a sin. The inscription contained secret knowledge from the Watchers—beings who had studied the movements of the sun, moon, and stars to find hidden signs in the sky. Fearing Noah's anger, Kainam kept this discovery a secret.

In the first year of the second week of the thirtieth jubilee, Kainam married Melka, the daughter of Madai, who was Japheth's son. Four years later, they had a son named Shelah, who said, "I have truly been sent." In the fifth week of the thirty-first jubilee, Shelah married Mu'ak, the daughter of Kesed, his father's brother. Five years later, she gave birth to a son named Eber. Later, in the thirty-second jubilee, during the seventh week, Eber married 'Azûrâd, the daughter of Nebrod. In the sixth year of that week, they had a son named Peleg. He was given this name because, during his lifetime, Noah's descendants started dividing the land among themselves. They first made a secret agreement on how to split it, then told Noah about their decision.

At the start of the thirty-third jubilee, in the first year of the first week, the earth was officially divided into three regions: one for Shem, one for Ham, and one for Japheth. This division was made under the guidance of a messenger sent by God. Noah's descendants and their families gathered to decide the borders of each territory.

Shem's land was in the middle of the earth, and he and his descendants were meant to live there forever. His territory started at the middle of the Rafa mountains, where the Tina River begins. It stretched west along the river to the waters of the deep and flowed into the Sea of Me'at, then continued toward the Great Sea. Japheth's land was to the north of Shem's, while Ham's land lay to the south.

Shem's territory extended south to Karaso and followed the western coastline of the Great Sea until it reached the mouth of the Egyptian Sea. From there, it went down the shores of the Great Sea to 'Afra, then reached the Gihon River. It followed the southern banks of the river eastward, passing just below the Garden of Eden. His land included Eden, the eastern regions, and all the way back to the Rafa mountains and the Tina River. This area, including the lands of Eden, was Shem's permanent inheritance.

Noah was pleased with this division because it matched the blessing he had spoken: "Blessed be the Lord God of Shem, and may the Lord dwell in his land." He knew that the Garden of Eden was the holiest place on earth, where God's presence was. He also knew that Mount Sinai, in the wilderness, and Mount Zion, at the center of the earth, were sacred places made to align with each other. Noah praised God for His wisdom and power and recognized the special blessing given to Shem and his descendants. Shem's land included Eden, the Red Sea region, India, the lands surrounding the Red Sea, and mountain ranges like Bashan, Lebanon, Sanir, Amana, Asshur, and Elam, as well as the Ararat region. This large territory was rich and full of resources.

Ham's land was located south of the Garden of Eden, beyond the Gihon River. His territory stretched to the fiery mountains and reached the 'Atel Sea. It extended west to the sea of Ma'uk, where many lost things eventually ended up. From there, it moved north to Gadir, followed the Great Sea's coastline, and circled back to the Gihon River, returning near the Garden of Eden. This land was given to Ham and his descendants as their permanent inheritance.

Japheth's land was north of the Tina River and covered the northern and northeastern regions, stretching all the way to the land of Gog. It extended eastward to the Qelt mountains and the sea of Ma'uk, then curved toward Gadir and beyond. To the west, it reached

Fara, then turned toward 'Aferag before extending east to the Sea of Me'at. Japheth's land continued northeast, reaching the Tina River and the Rafa mountains. His territory was vast and included five large islands. Japheth's land was known for its cold climate, while Ham's land was hot, and Shem's land had a balanced, temperate climate in between.

Chapter IX.

Ham divided his land among his sons. Cush received the first portion, which was in the eastern region. To the west of Cush was Mizraim's land, followed by Put's territory even farther west. The farthest west, along the seacoast, was the land given to Canaan.

Shem also divided his inheritance among his sons. Elam received the first portion, which covered the land east of the Tigris River and stretched further east, including all of India. His territory also included the coastline along the Red Sea, the waters of Dedan, the mountains of Mebri and Ela, the land of Susan, and the regions near Phatnak, reaching the Red Sea and the Tina River.

The second portion was given to Asshur, which included all the land of Assyria, along with the cities of Nineveh and Shinar, as well as the borderlands of India. This region followed the path of the major rivers in the area.

The third portion went to Arpachshad, covering the entire land of the Chaldeans, east of the Euphrates River, near the Red Sea. His inheritance also included the desert lands near the sea's mouth facing Egypt, along with Lebanon, Sanir, and 'Amana, extending to the Euphrates River.

The fourth portion was given to Aram. His land was located between the Tigris and Euphrates Rivers, north of the Chaldean territory, and stretched to the Asshurite mountains and the land of

'Arara.

Lud, Shem's fifth son, received the mountains of Asshur and the surrounding lands. His territory extended to the Great Sea and stretched eastward, toward the land of his brother Asshur.

Japheth also distributed his land among his sons. Gomer received the first portion, which stretched eastward from the northern region to the Tina River. To the north, Magog's land included the inner northern territories, extending toward the Sea of Me'at.

Madai's land was located west of his brothers' territories and included islands and coastal regions.

Javan, who received the fourth portion, was given all the islands and lands that bordered Lud's territory.

Tubal received the fifth portion, which covered the central landmass bordering Lud. His land extended to the second landmass and stretched beyond into the third section of land.

Meshech was given the sixth portion, which extended beyond the third landmass and reached the eastern border of Gadir.

Tiras received the seventh portion, which included four large islands located in the sea. These islands extended toward the border of Ham's land. The Kamaturi Islands were also assigned to the descendants of Arpachshad as part of their inheritance.

Noah's sons divided the earth among their descendants while Noah was still alive. He made them take an oath, binding them under a curse so that no one would take land that was not given to them.

They all agreed and swore, saying, "So be it, so be it," for themselves and for their future generations. This agreement was meant to last forever, until the day of judgment. On that day, the Lord will judge all nations with fire and the sword for their sins, their corruption, and the evil they have spread across the earth.

Chapter X.

During the third week of that jubilee, unclean spirits began leading Noah's descendants astray, causing confusion and destruction. Noah's sons came to him, troubled by how these demons were deceiving, blinding, and even killing their children.

Noah prayed to the Lord, saying:

"God of all spirits, You have shown mercy to me and my family, saving us from the flood and sparing us from the fate of the wicked. Your kindness and compassion have been great.

Please continue Your grace upon my sons. Do not let these evil spirits take control of them or lead them to destruction. Bless me and my children so that we may grow in number and fill the earth. You know how the Watchers, the ancestors of these spirits, acted in my time. Bind these spirits and keep them in the place of punishment so they cannot harm Your faithful servants. They were created for destruction and should not have power over those who seek righteousness.

Do not let them rule over the living or the righteous, for only You have authority over all spirits. Do not allow them to overpower those who choose to walk in Your ways forever."

The Lord heard Noah's prayer and ordered us to bind the spirits. Then, Mastema, the leader of these spirits, stepped forward and said:

"Lord, Creator of all, allow some of them to remain with me to carry out my commands. Without them, I cannot fulfill my purpose of testing and corrupting mankind, for their wickedness is great."

The Lord replied:

"I will allow one-tenth of them to remain with you, but the other nine-tenths must be bound in the place of punishment."

The Lord then commanded one of us to teach Noah how to heal diseases, knowing that mankind would continue to act wickedly. We obeyed, binding the evil spirits and leaving only one-tenth under Mastema's control on earth. We also taught Noah how to use herbs from the earth to heal sickness and resist the temptations brought by these spirits. Noah wrote everything down in a book, recording all the remedies and instructions we gave him. This knowledge helped protect his descendants from harm.

Noah passed these teachings to his eldest son, Shem, whom he loved most. After living a righteous life, Noah died and was buried on Mount Lubar in the land of Ararat. He lived for 950 years, completing 19 jubilees, two weeks, and five years. Among all men, he was the most righteous, second only to Enoch, who served as a testimony for future generations until the day of judgment.

During the thirty-third jubilee, in the first year of the second week, Peleg married Lomna, the daughter of Sina'ar. In the fourth year, she gave birth to a son, Reu. Peleg said, "Look, the people have become wicked, for they have begun to build a city and a tower in the land of Shinar."

The people had moved away from the land of Ararat and settled in Shinar. There, they decided to build a great city and a tower that would reach the heavens. They said, "Let's build a tower so high that we will make a name for ourselves." They worked for 43 years, making bricks and using asphalt from the sea as mortar. The tower rose to an incredible height of 5,433 cubits and 2 palms, with its walls extending 13 and 30 stades.

The Lord said, "They are one people with a single language, and now nothing they plan will be impossible for them. Let us go down and confuse their language so they will not understand one another."

We went down with the Lord to see the city and the tower. He confused their language, making them unable to continue their work. The land of Shinar was called Babel because the Lord scattered the people across the earth. A powerful wind knocked the tower down, and its ruins remained between Asshur and Babylon in Shinar. The people were dispersed in the first year of the thirty-fourth jubilee.

Ham and his sons moved to their assigned land in the south. However, Canaan saw the land of Lebanon, from the river of Egypt, and found it desirable. Instead of settling in the land given to him by the sea, he chose to stay in Lebanon, east and west of the Jordan and along the coast.

Ham, along with Cush and Mizraim, warned Canaan:

"You have taken land that does not belong to you. This land was given to Shem and his descendants. If you stay here, you and your children will be cursed and driven out for your disobedience. Do not live in Shem's territory, for it was assigned to him and his descendants by God."

But Canaan ignored them and remained in Lebanon, from Hamath to the borders of Egypt, along with his sons. That is why the land became known as Canaan.

Japheth and his sons moved to their rightful land by the sea. However, Madai did not like his assigned territory. He asked Ham, Asshur, and Arpachshad for a piece of their land. He settled in Media, near his brother-in-law, and named the land after himself. The name Media has remained ever since.

Chapter XI.

During the thirty-fifth jubilee, in the third week of its first year, Reu married a woman named Ôrâ, the daughter of Ûr, who was the son of Kesed. In the seventh year of that week, she gave birth to a son named Serôh. Around this time, Noah's descendants started fighting among themselves. They captured and killed each other, spilling human blood across the land. Some even began eating blood. They built fortified cities, walls, and towers. People became proud, forming the first kingdoms and waging wars—city against city, nation against nation. Weapons were created, children were trained in warfare, and men started capturing others to sell as slaves.

Ûr, the son of Kesed, built the city of Ara in the land of the Chaldees, naming it after himself and his father. The people made molten idols and began worshiping them. They carved statues and created impure images. Evil spirits led them further into sin, and Mastêmâ, the ruler of these spirits, worked tirelessly to spread corruption. He sent his demons to commit all kinds of wickedness, violence, and bloodshed. Because of the widespread sin during this time, Serôh was later called Serug, a name reflecting the transgressions of that era.

Serug grew up in Ur of the Chaldees, near his wife's mother's family. Sadly, he also worshiped idols. During the thirty-sixth jubilee, in the first year of the fifth week, Serug married Melka, the daughter of Kaber, who was the daughter of his father's brother. In the first year of that week, she gave birth to a son named Nahor. Nahor was raised in Ur of the Chaldees, where his father taught him the ways of their people, including astrology and fortune-telling.

During the thirty-seventh jubilee, in the first year of the sixth week, Nahor married 'Ijaska, the daughter of Nestag from the Chaldees. In the seventh year of that week, they had a son named Terah. During this time, Mastêmâ sent flocks of ravens and birds to destroy crops and steal seeds before they could be planted. As a result, Terah's name was given to symbolize the hardship caused by the ravens. The land became barren, and people struggled to gather enough food from their harvests.

In the thirty-ninth jubilee, during the first year of the second week, Terah married Edna, the daughter of Abram, who was his father's sister. In the seventh year of that week, she gave birth to a son, whom they named Abram, after his grandfather, who had died before his birth.

As Abram grew older, he began to see the mistakes and wickedness of the world. He noticed that the people around him worshiped idols and practiced unclean rituals. His father taught him how to write, but by the time he was 14, Abram chose to distance himself from his father's ways to avoid idol worship. Instead, he prayed to the Creator of all things, asking for guidance to live a pure and righteous life.

When the season for planting arrived, people gathered in the fields to guard their seeds from the ravens. Although he was still a boy, Abram went with them. When a massive flock of ravens came to eat the seeds, Abram ran toward them and shouted, "Do not come down! Go back to where you came from!" Miraculously, the ravens turned away. That day, Abram drove the ravens away seventy times, and not a single bird remained in the land where he lived.

The people were amazed and word of Abram spread throughout the Chaldees. That year, farmers sought his help, and he guided them during the planting season. With his assistance, they successfully sowed all their seeds. That year, the harvest was abundant, and the people rejoiced.

In the first year of the fifth week, Abram invented a new tool for oxen plows to prevent ravens from stealing the seeds. He designed a wooden container that attached to the plow, allowing seeds to drop directly into the soil as the oxen moved. This clever invention prevented birds from eating the seeds before they could grow. The people quickly adopted Abram's design, and soon, they were able to plant and harvest without fear of the birds. Abram's wisdom and leadership brought prosperity and peace to the land.

Chapter XII.

During the sixth week, in the seventh year, Abram spoke to his father, Terah, and said, "Father!"

Terah answered, "I am here, my son."

Abram asked, "Why do you worship and bow down to these idols? They have no life, no spirit, and they only lead people astray. Why put your trust in them?

Worship the God of heaven, the One who sends rain and dew, who controls everything on earth, and who created all things with His word. All life comes from Him.

Why believe in lifeless statues made by human hands? You carry them on your shoulders, yet they cannot help you. They bring shame to those who make them and deceive those who worship them. Do not follow them."

Terah replied, "I know what you are saying is true, my son, but what can I do? The people here force me to serve these idols. If I speak against them, they will kill me because they are committed to worshiping them. Stay silent, my son, or they will kill you too."

Abram shared these thoughts with his brothers, but they became angry with him. So he remained quiet.

During the fortieth jubilee, in the second week and the seventh year, Abram married Sarai, his father's daughter, and she became his wife. His brother Haran also married in the third year of the third week, and in the seventh year, his wife gave birth to a son named Lot. Their other brother, Nahor, also took a wife.

When Abram was sixty years old, in the fourth week and the fourth year, he woke up in the middle of the night and set fire to the house of idols, burning everything inside. No one knew he had done it. When the people awoke, they rushed to save their gods from the fire. Haran tried to rescue them, but the flames overtook him, and he died in Ur of the Chaldees in front of his father, Terah. They buried him there.

After this, Terah left Ur with his sons and traveled toward the land of Lebanon and Canaan. He settled in Haran, where Abram stayed with him for fourteen years.

During the sixth week, in the fifth year, Abram stayed awake on the new moon of the seventh month to observe the stars and predict the coming rainfall. As he watched, he thought, "All the movements of the stars, the moon, and the sun are in the hands of the Lord. Why am I searching for answers in them?

If God wills, He sends rain in the morning and evening. If He chooses, He withholds it. Everything happens according to His will."

That night, Abram prayed, saying, "My God, the Most High, You are my only God, and I have chosen to follow You alone. You created everything, and all that exists is the work of Your hands.

Protect me from the evil spirits that mislead people so that I will not be led away from You. Strengthen me and my descendants forever, so we will never turn from Your ways."

Then he asked, "Should I return to Ur of the Chaldees, where they are calling me back? Or should I stay here? Guide me, O God, and make my path clear so that I do not follow my own desires."

After Abram finished praying, the word of the Lord came to him, saying:

"Leave your land, your people, and your father's house, and go to the land I will show you. I will make you into a great and mighty nation.

I will bless you and make your name great, and through you, all the families of the earth will be blessed. I will bless those who bless you and curse those who curse you.

Do not be afraid, for I will be your God forever, for you and your descendants through all generations."

The Lord also commanded, "Open his mouth and ears so that he may understand and speak the language I have given." This was the original language of creation, which had not been spoken since the Tower of Babel. At that moment, Abram's mouth, ears, and lips were opened, and the Lord spoke to him in Hebrew.

Abram then took the writings of his ancestors, which were written in Hebrew, and copied them, studying them carefully. The Lord helped him understand what he could not comprehend, and Abram spent six months of the year studying these writings during the rainy season.

In the seventh year of the sixth week, Abram told his father that he planned to leave Haran to visit the land of Canaan and return later.

Terah said to him, "Go in peace. May the eternal God guide you, protect you from harm, and grant you grace, mercy, and favor in the eyes of those you meet. May no one have the power to hurt you. Go safely.

If you find a land that pleases you, take me with you, and take Lot, the son of your brother Haran, as your own. But leave Nahor, your brother, with me. When you return safely, we will all go with you."

Chapter XIII.

Abram left Haran with his wife Sarai and his nephew Lot, traveling to the land of Canaan. On the way, they passed through Asshur and arrived at Shechem, where they settled near a large oak tree. The land was beautiful, stretching from Hamath to the great oak. Then, the Lord appeared to Abram and said, "I will give this land to you and your future family." Abram built an altar there and made a burnt offering to God.

After that, Abram moved to a mountain between Bethel in the west and Ai in the east, where he set up his tent. He saw that the land was rich and full of life, with vineyards, fig trees, pomegranates, olive trees, cedars, date palms, and many other plants. Water flowed from the mountains, making the land fertile. Abram gave thanks to the Lord, who had guided him safely from Ur of the Chaldees to this land of blessings.

In the first year of the seventh week, on the first day of the first month, Abram built an altar on the mountain and prayed, saying, "You are the eternal God, and You are my God." He offered a burnt sacrifice, asking God to always be with him. Then, Abram traveled south and arrived in Hebron, where a city was being built. He stayed there for two years before moving further south to Bealoth. While he was there, a famine spread across the land.

In the third year of that time, Abram went to Egypt and stayed for five years. During his stay, Pharaoh took Sarai into his palace. But the Lord sent terrible plagues upon Pharaoh and his household because of her. As a result, Abram became very wealthy, gaining many sheep, cattle, donkeys, horses, camels, servants, silver, and gold. His nephew Lot also became rich.

Pharaoh returned Sarai to Abram and sent them out of Egypt.

Abram traveled back to the place where he had first set up his tent, between Bethel and Ai, near the altar he had built before. There, he thanked the Lord for bringing him back safely. In the forty-first jubilee, during the third year of the first week, Abram made another burnt offering at the altar and prayed, saying, "You are the Most High God, and You are my God forever."

In the fourth year of that time, Lot separated from Abram and moved to Sodom, where the people were extremely sinful. Abram was saddened by Lot's choice, especially since he had no children of his own. Later that year, after Lot was taken captive, the Lord spoke to Abram and said, "Look around in every direction—north, south, east, and west. I will give all this land to you and your descendants forever. Your family will be as numerous as the dust of the earth. Walk through the land, for it will belong to you and your future generations."

Abram then moved to Hebron and settled there. That same year, Chedorlaomer, the king of Elam, joined forces with Amraphel, the king of Shinar, Arioch, the king of Sellasar, and Tergal, the king of the nations. Together, they attacked the king of Gomorrah. The king of Sodom fled, and many people were wounded in the Siddim Valley near the Salt Sea. The invading kings took over Sodom, Adam, and Zeboim, capturing Lot and taking all his belongings to Dan.

One of the survivors escaped and told Abram what had happened to Lot. Abram quickly gathered his trained servants, armed them, and went after the enemy. He successfully rescued Lot, recovered his possessions, and brought back the people who had been taken. After the victory, Abram gave one-tenth of the recovered goods to the Lord, establishing a lasting rule that a tenth of all produce—grain, wine, oil, cattle, and sheep—should be given to the priests who serve before God.

The king of Sodom then approached Abram, bowed before him, and said, "Lord Abram, keep the goods for yourself, but return the people you rescued to me." Abram replied, "I swear to the Most High God that I will take nothing from you—not even a thread or a sandal strap—so that you cannot say, 'I made Abram rich.' The only things I will take are what my men have already eaten and the share that belongs to Aner, Eschol, and Mamre, who helped me. They will receive their portion."

Chapter XIV.

In the fourth year of that time, on the first day of the third month, the Lord spoke to Abram in a vision, saying, "Do not be afraid, Abram. I am your protector, and your reward will be very great." Abram replied, "Lord, what can You give me if I still have no children? The one who will inherit everything I own is Eliezer of Damascus, the son of my servant Maseq. You have not given me any children of my own."

The Lord answered, "Eliezer will not be your heir. You will have a child of your own, and he will inherit everything." Then, the Lord took Abram outside and said, "Look up at the sky and try to count the stars, if you can." As Abram looked at the endless stars above him, the Lord said, "That is how many descendants you will have." Abram believed what the Lord had promised, and because of his faith, God considered him righteous. Then the Lord said, "I am the one who brought you out of Ur of the Chaldees to give you this land as your inheritance forever. I will be your God and the God of your descendants."

Abram asked, "Lord, how can I be sure that I will inherit this land?" The Lord told him, "Bring Me a three-year-old cow, a three-year-old goat, a three-year-old ram, a turtledove, and a young pigeon."

Abram did as the Lord commanded. In the middle of the month, he stayed near the oak trees of Mamre, close to Hebron. There, he built an altar, sacrificed the animals, and poured their blood on it. He cut the animals in half and placed the pieces across from each other, but he did not divide the birds. Then, large birds came down, trying to eat the sacrifices, but Abram chased them away.

As the sun began to set, Abram fell into a deep sleep. A heavy darkness surrounded him, filling him with fear. Then, the Lord spoke, saying, "Know for certain that your descendants will live in a land that is not their own. They will be enslaved and mistreated for 400 years. But I will punish the nation that enslaves them, and in the end, they will leave with many possessions. You, however, will live in peace and grow old before being buried. In the fourth generation, your descendants will return to this land, for the sins of the Amorites are not yet complete."

When Abram woke up, the sun had already set. He saw a blazing fire and a cloud of smoke pass between the divided pieces of the sacrifice. That day, the Lord made a covenant with Abram, saying, "I will give this land to your descendants—from the river of Egypt to the great river, the Euphrates. This land belongs to the Kenites, Kenizzites, Kadmonites, Hittites, Perizzites, Rephaim, Amorites, Canaanites, Girgashites, and Jebusites."

Abram completed the offerings, including the birds and their grain and drink offerings, which were burned in the fire. The Lord sealed His covenant with Abram on that day, just as He had done with Noah in this same month. Abram renewed the festival and the practice as a tradition for himself and his future generations.

Overjoyed, Abram shared everything that had happened with his wife, Sarai. He believed fully in God's promise that he would have many descendants. However, Sarai still had not been able to conceive.

She told Abram, "Take my Egyptian maid, Hagar, as your wife. Maybe I can have children through her." Abram listened to Sarai and agreed. She gave Hagar to Abram as his wife. He was with Hagar, and she became pregnant and gave birth to a son. Abram named him Ishmael. This happened in the fifth year of that time, when Abram was 86 years old.

Chapter XV.

In the fifth year of the fourth week of this special time, during the third month, Abraham celebrated the Festival of First Fruits from the grain harvest. He made offerings to God on the altar, presenting the first portion of his crops. He sacrificed a young cow, a goat, and a sheep as burnt offerings, along with grain and drink offerings, and sprinkled frankincense on the altar.

The Lord appeared to Abraham and said, "I am God Almighty. Follow My ways and live with honesty and integrity. I will make a covenant with you and greatly increase your descendants." Abraham bowed down to the ground, and God continued, "My covenant is with you, and you will become the ancestor of many nations. From now on, your name will no longer be Abram but Abraham, because I have made you the father of many nations. I will bless you abundantly, and your family will grow. Nations and kings will come from you. This covenant will last forever between Me and your descendants. I will be your God and the God of your future generations. I will give you and your family the land of Canaan, where you now live as foreigners, as a permanent possession, and I will be their God."

Then God gave Abraham instructions, saying, "You and your descendants must keep My covenant for all generations. Every male among you must be circumcised as a sign of this agreement. On the eighth day after birth, every male must be circumcised, whether he was born in your household or bought from a foreigner. This will be a permanent mark of the covenant between us. Any male who is not circumcised on the eighth day has broken My covenant and will be cut off from his people."

God also said, "As for your wife, she will no longer be called Sarai. Her name will now be Sarah. I will bless her, and she will give birth to a son. She will be the mother of nations, and kings will come from her." Abraham bowed down and laughed to himself, thinking, "How can a hundred-year-old man have a child? Can Sarah, at ninety years old, give birth?" Then Abraham said to God, "If only Ishmael could receive Your blessing!"

But God replied, "No, Sarah will give you a son, and you will name him Isaac. My everlasting covenant will be with him and his descendants. As for Ishmael, I have heard your request. I will bless him, make him fruitful, and give him many descendants. He will become the father of twelve rulers, and I will make him into a great nation. But My covenant will be with Isaac, whom Sarah will give birth to at this time next year."

After God finished speaking, He left. Abraham immediately obeyed God's command. That same day, he circumcised Ishmael, every male in his household—whether born there or bought—and himself. Every male in his household was circumcised that day as a sign of the covenant. The commandment to circumcise boys on the eighth day was recorded as a permanent law on the heavenly tablets. Anyone who failed to do this would be removed from the covenant and separated from God's people.

All the angels who serve in God's presence were created to worship Him, and God chose Israel as His special people even before they existed. He declared that Israel would always be with Him and

His holy angels. The children of Israel were commanded to keep this covenant forever. If they remained faithful, they would never be removed from their land. This law was given as an everlasting command.

Even though Ishmael and Esau were Abraham's descendants, God did not choose them to be near Him. He chose Israel, set them apart, and made them His people, different from all other nations. While every nation belongs to God, He placed spiritual rulers over them for guidance. But He reserved Israel for Himself, without any intermediaries. He alone is their God, leading and protecting them forever.

However, I must warn you that the children of Israel will not keep this covenant. They will fail to circumcise their sons, ignoring this eternal law, and will leave them uncircumcised as they were born. This will anger the Lord because they will have abandoned His covenant, rejected His laws, and followed the customs of other nations. Their rebellion and disrespect will lead to their exile, and they will be cast out of their land. There will be no forgiveness for them because they will have broken this everlasting covenant.

Chapter XVI.

At the start of the fourth month, under the shade of the large oak at Mamre, we visited Abraham. We told him that his wife, Sarah, would have a son. Sarah, listening nearby, laughed quietly to herself because she didn't believe it was possible. We reassured her and told her not to be afraid, though she denied laughing. Still, we revealed that her son's name, Isaac, had already been written in the heavenly records. We promised that when we returned at the right time, she would be expecting a child.

During that same time, the Lord brought judgment upon Sodom,

Gomorrah, Zeboim, and the surrounding areas near the Jordan. Fire and sulfur rained down, destroying them completely. To this day, those cities remain in ruins. Their wickedness had reached its limit, as they had corrupted themselves and spread evil everywhere. Just as Sodom was punished, so will any place that follows in their footsteps.

However, God showed mercy to Lot because of Abraham. He rescued Lot from the destruction, but even after escaping, Lot and his daughters committed a terrible sin—something unheard of since the time of Adam. Their actions were recorded as a serious wrongdoing in the heavenly records. Because of this, it was decided that Lot's descendants would not survive. Just like Sodom, his family line would be cut off. Their judgment is certain, and when the time comes, none of them will remain.

That same month, Abraham left Hebron and traveled toward the area between Kadesh and Shur. He settled in the mountains near Gerar. By the middle of the fifth month, he moved again, this time to the Well of the Oath. Then, in the middle of the sixth month, as promised, the Lord visited Sarah, and she became pregnant. Just as God had said, Sarah gave birth to a son in the third month, on the day of the Festival of First Fruits of the Harvest. And so, Isaac was born, fulfilling God's promise.

We told Abraham that while his other sons would be connected to different nations, Isaac's descendants would be set apart as a holy people, chosen by God. His family line would belong to the Most High, forming a special kingdom and priesthood devoted to serving Him. After delivering this message, we left Abraham and went to Sarah, sharing the same words with her. Both Abraham and Sarah rejoiced deeply at this news.

Abraham built an altar to honor the Lord, who had protected him and blessed him with great joy, even though he was living in a foreign land. At the altar near the Well of the Oath, he held a seven-day festival of celebration. During this time, he made temporary shelters for himself and his household. This was the first time the Feast of Tabernacles was observed on earth. Every day for seven days, Abraham made offerings to the Lord, including two oxen, two rams, seven sheep, and one male goat as a sin offering, asking for forgiveness for himself and his descendants.

Along with these offerings, he also gave thanksgiving sacrifices, which included seven rams, seven goats, seven sheep, and seven male goats, as well as grain and drink offerings. He burned all the fat on the altar as a pleasing aroma to the Lord. Every morning and evening, Abraham burned incense made from a special blend of spices: frankincense, galbanum, stacte, nard, myrrh, costus, and other fragrant spices. He combined them in equal amounts to create a pure and sweet-smelling incense for God.

For the full seven days, Abraham joyfully celebrated the festival with complete devotion. His entire household took part in the observance, but no outsiders or uncircumcised people were allowed to join. Abraham praised God, giving thanks for creating him and guiding him according to His divine plan. God already knew that Abraham's descendants would follow the path of righteousness, and from his family would come a holy people who reflected His goodness.

With joy and respect, Abraham honored God and named this celebration the "Festival of the Lord," a time of rejoicing that pleased the Most High. We blessed Abraham and his descendants forever because he followed the festival exactly as it was written in the heavenly records. Because of this, it was decided in the heavenly writings that Israel would celebrate the Feast of Tabernacles every year for seven days in the seventh month. This was to be a lasting commandment for all generations.

This festival would never be forgotten. It was established that Israel must observe it every year. They were instructed to live in temporary shelters, wear wreaths on their heads, and take leafy branches and willow branches from the streams. Abraham gathered palm branches and beautiful fruits, and each morning, he walked around the altar seven times, giving thanks and praising God with great joy for all that He had done.

Chapter XVII.

In the first year of the fifth week of that special time, Isaac was weaned, and Abraham held a great feast in the third month to celebrate. Ishmael, the son of Hagar the Egyptian, stood beside Abraham, and Abraham felt great joy. He praised God for giving him sons and not leaving him without children. He also remembered the promise God had made to him when Lot had separated from him, and his heart was full of gratitude as he gave thanks to the Creator.

However, Sarah saw Ishmael playing and dancing while Abraham was celebrating, and she became jealous. She said to Abraham, "Get rid of this slave woman and her son. Her son will not share in the inheritance with my son, Isaac." Abraham was deeply troubled by this because it involved both his servant and his son.

But God spoke to Abraham and said, "Do not be distressed about the boy or the maidservant. Listen to what Sarah is saying, because your descendants will come through Isaac. But do not worry—I will also make a great nation from the son of the slave woman because he is your child too."

The next morning, Abraham got up early, took some bread and a skin of water, placed them on Hagar's shoulder along with her son, and sent them away.

Hagar wandered in the wilderness of Beersheba. When the water ran out, the child became weak and collapsed. She placed him under the shade of an olive tree and walked a short distance away, saying, "I cannot bear to watch my child die." She sat down and wept.

Then an angel of God appeared to her and said, "Hagar, why are you crying? Get up, take the child, and hold him, for God has heard your cries and seen your child's suffering." Suddenly, Hagar saw a well of water. She quickly filled the water skin and gave the boy a drink. Then, they continued on to the wilderness of Paran. The boy grew up and became an excellent archer, and God was with him. Later, Hagar found him a wife from Egypt, and she gave birth to a son. He named him Nebaioth, saying, "The Lord was near to me when I called upon Him."

In the first year of the seventh week, on the twelfth day of the first month, voices from heaven spoke about Abraham. They declared that he had been faithful in everything God had commanded him, that he truly loved the Lord, and that he had proven his loyalty in every test.

Then Mastêmâ, the adversary, came before God and said, "Abraham may love You, but he loves his son Isaac even more. Command him to offer Isaac as a burnt sacrifice on the altar, and then You will see if he is truly obedient. Then You will know if he is faithful in everything."

But the Lord already knew Abraham's heart and that he was strong through every test. God had already tested him when He called him to leave his homeland, when he faced famine, when he encountered the riches of kings, and when his wife was taken from him. God tested him when He gave him the covenant of circumcision and again when he had to send Ishmael and Hagar away. Through all these trials, Abraham remained faithful and patient. He never hesitated to follow God's instructions because he loved the Lord and was completely devoted to Him.

Chapter XVIII.

One day, God called out, "Abraham, Abraham!" and Abraham answered, "Here I am." Then God said, "Take your son, Isaac—the one you love—and go to the high mountains. There, on a mountain that I will show you, offer him as a burnt sacrifice."

Early the next morning, Abraham got up, saddled his donkey, and took two of his servants along with Isaac. He cut the wood for the burnt offering and set out for the place God had told him about. After traveling for three days, he saw the mountain in the distance.

When they arrived at a well, Abraham said to his servants, "Stay here with the donkey. Isaac and I will go up the mountain to worship. After we have worshiped, we will come back to you."

Abraham took the wood for the offering and placed it on Isaac's shoulders. He himself carried the fire and the knife as they walked together toward the mountain.

On the way, Isaac spoke to his father, saying, "Father?" Abraham replied, "Yes, my son?" Isaac asked, "We have the fire and the wood, but where is the lamb for the burnt offering?"

Abraham answered, "God will provide the lamb for the sacrifice, my son." And the two of them continued on.

When they reached the place God had chosen, Abraham built an altar and arranged the wood on it. Then, he tied up Isaac and placed him on the altar, on top of the wood. Abraham reached for the knife and was about to sacrifice his son.

At that moment, I was there, along with Mastêmâ, and we heard God call out, "Abraham, Abraham!" Abraham quickly responded, "Here I am." God said, "Do not harm the boy. Now I know that you truly respect and obey Me, because you were willing to offer your son, your only son."

Then, I called out to Abraham from heaven again, saying, "Abraham, Abraham!" He answered, "Here I am." I told him, "Do not lay a hand on the boy. You have shown your deep trust in God by not holding back your son from Him."

Mastêmâ was left in shame. Abraham looked up and saw a ram caught by its horns in the bushes. He went over, took the ram, and offered it as a sacrifice in place of Isaac.

Abraham named that place "The Lord Will Provide," and even today, people say, "On the mountain of the Lord, it will be provided."

Then God called out to Abraham again from heaven and said, "I swear by Myself," declares the Lord, "Because you have done this and did not hold back your beloved son, I will bless you greatly. I will make your descendants as numerous as the stars in the sky and the grains of sand on the shore. They will conquer the cities of their enemies, and through your descendants, all the nations of the earth will be blessed, because you have obeyed My voice. You have proven your faithfulness in everything I have asked of you. Now go in peace."

Abraham returned to his servants, and together they traveled back to Beersheba, where he settled near the Well of the Oath.

From that time on, Abraham celebrated this event every year for seven days with great joy. He named it the Festival of the Lord, remembering the seven days of his journey and safe return.

It is recorded in the heavenly writings that Israel and its future generations must observe this festival every year for seven days, celebrating with joy.

Chapter XIX.

In the first year of the first week of the forty-second jubilee, Abraham returned and settled near Hebron, in a place called Kirjath Arba. He lived there for fourteen years.

In the first year of the third week, Sarah passed away in Hebron. Abraham mourned for her and arranged her burial. During this time, he was tested to see if he would remain patient and free from anger, and he passed the test by staying calm and composed.

He kindly approached the sons of Heth and asked for a burial place for his wife. The Lord made them favor him, and they treated him with respect. Abraham asked for the field that contained the cave near Mamre, also known as Hebron. They agreed to give it to him for four hundred pieces of silver. Even though they offered it as a gift, Abraham insisted on paying the full price. After completing the purchase, he bowed before them twice and buried Sarah in the cave.

Sarah lived for 127 years—two jubilees, four weeks, and one year in total. Her passing was Abraham's tenth test of faith, and he remained faithful and patient. Even though God had promised him the land, he still humbly asked for a burial site instead of questioning God's promise. Because of his faith, he was honored in the heavenly records as a friend of God.

In the fourth year of that time, Abraham arranged a marriage for Isaac. He chose Rebecca, the daughter of Bethuel, who was the son of Nahor, Abraham's brother. Around the same time, Abraham also married another wife, Keturah, who came from the daughters of his servants. Hagar had already passed away before Sarah. Over the next fourteen years, Keturah gave birth to six sons: Zimram, Jokshan, Medan, Midian, Ishbak, and Shuah.

In the second year of the sixth week, Rebecca gave birth to twin sons, Jacob and Esau. Jacob was quiet and righteous, living peacefully in tents, while Esau was wild, spending his time hunting in the fields and becoming skilled in battle. Abraham loved Jacob, but Isaac favored Esau.

As Abraham watched Esau's behavior, he realized that Jacob, not Esau, would carry on his name and legacy. He called Rebecca, knowing she loved Jacob more than Esau, and said:

"My daughter, take great care of Jacob,
For he will inherit my place on this earth.
He will bring blessings to all people
And will bring honor to the line of Shem.

The Lord has chosen him to be His own people,
Set apart from all the nations of the earth.
Though Isaac loves Esau more,
You love Jacob, and I ask you to care for him even more.

Let your love for him guide your actions,
For he will bring blessings to us
And to all future generations.

Be strong and take joy in your son Jacob,
For I have loved him more than all my children.

He is blessed forever,
And his descendants will fill the earth.
If a man could count the grains of sand on the earth,
Jacob's descendants would be just as many.

All the blessings God has given me
Will belong to Jacob and his descendants forever.

Through his family, my name and the names of my ancestors—
Shem, Noah, Enoch, Mahalalel, Enos,
Seth, and Adam—will be honored.

These blessings will uphold the heavens,
Strengthen the earth,
And renew the stars above."

Then Abraham called Jacob to stand before Rebecca and kissed him. He blessed him, saying:

"My beloved son Jacob, whom I cherish,
May God bless you from the heavens above.
May He give you all the blessings He gave to Adam, Enoch,
    Noah, and Shem.
May all the promises He made to me and our family be fulfilled
    in you,
And may those blessings last forever, as long as the heavens
    remain above the earth.
May no spirit of Mastêmâ have power over you or your
    descendants,
To lead you away from the Lord your God,
From this day forward and forever.

May the Lord be your Father,
And may you be His firstborn son,
A blessing to His people for all time.

Go in peace, my son."

After this, Abraham and Jacob spent time together, and Rebecca loved Jacob with all her heart, far more than Esau. However, Isaac continued to favor Esau, loving him more than Jacob.

Chapter XX.

In the forty-second jubilee, during the first year of the seventh week, Abraham gathered his family together. He called Ishmael and his twelve sons, Isaac and his two sons, and the six sons of Keturah along with their children.

Abraham taught them to follow the ways of the Lord, to live with honesty and kindness, and to treat others with fairness and justice. He urged them to stay true to God's commandments, never turning away from them. He warned them to avoid all kinds of immoral and impure behavior and to make sure such actions did not take place in their families or communities.

He stressed that if any woman or girl among them committed an immoral act, she should be punished, and no one should desire her or seek her out. He also warned them not to marry the daughters of Canaan because the people of Canaan would one day be removed from the land.

He reminded them about the punishment that came upon the giants and the people of Sodom. He described how they were destroyed because of their wickedness, their sins, and the corruption they spread.

Stay away from sin and anything unclean,
And always choose to do what is right.
Do not bring shame to our family,
Or disgrace to your own lives.

Do not let your children suffer violence,
Or bring a curse upon yourselves like Sodom,
Or have your descendants punished like the people of
Gomorrah.

My sons, I urge you to love the God of heaven,
And follow all His commandments.

Do not be led astray by false gods or their wicked ways.
Do not make idols for yourselves,
For they are useless and lifeless,
Created by human hands, and trusting in them is trusting in
  nothing.

Do not bow down to them or serve them,
But worship the Most High God and honor Him always.
Seek His favor and strive to do what is right,
So that He may be pleased with you, show you kindness,
And send rain for your fields in the morning and evening.

May He bless the work of your hands,
Bless your food and water,
Bless your children and your land,
And bless your animals and flocks.

You will be a blessing to the world,
And all nations will look to you with honor.
They will bless your children in my name,
So they too may receive the same blessings I have been given.

Abraham gave gifts to Ishmael and his sons, as well as to the sons
of Keturah. Then, he sent them away from Isaac, giving all that he
owned to Isaac.

Ishmael and his sons, along with the sons of Keturah and their
families, traveled together and settled in the lands stretching from
Paran to the entrance of Babylon, covering the eastern regions near

the desert. Over time, they intermarried and became known as the Arabs and the Ishmaelites.

Chapter XXI.

In the sixth year of the seventh week of this jubilee, Abraham called his son Isaac and said,"My son, I have grown old, and I do not know how much time I have left. I am now 175 years old, and my life has been full. Throughout the years, I have always remembered the Lord and done my best to follow His ways with all my heart. I have lived with honesty and integrity. I have rejected idols and those who worship them. My heart and soul have been fully devoted to obeying my Creator, for He is the one true God—holy, faithful, and completely just. He does not show favoritism or accept bribes. He judges fairly and will hold accountable those who break His laws or abandon His covenant.

Now, my son, follow His commandments, obey His instructions, and live according to His laws. Stay away from anything sinful or unclean, especially idol worship. Never eat the blood of any animal, whether from cattle, birds, or any other creature. If you offer a peace sacrifice, do it properly. Pour its blood on the altar and burn its fat along with fine flour mixed with oil and a drink offering. These will create a pleasing aroma to the Lord. When offering a thanksgiving sacrifice, burn the fat from the belly, the inner organs, the kidneys, and the fat near the loins and liver. Place these parts on the fire of the altar along with the meat and the drink offering as a sweet-smelling sacrifice to the Lord.

Eat the meat on the same day or the next, but never on the third day. If any remains until then, it is no longer acceptable and must not be eaten. Anyone who eats it on the third day commits a sin. I have read these instructions in the writings of our ancestors, in the words of Enoch and Noah. Also, always sprinkle salt on your offerings. The salt of the covenant must never be missing from any sacrifice you bring before the Lord.

When choosing wood for sacrifices, only use these kinds: cypress, bay, almond, fir, pine, cedar, savin, fig, olive, myrrh, laurel, or aspalathus. Pick wood that is strong, fresh, and looks good. Do not use wood that is cracked, dark, or damaged. Never use old wood, as it has lost its fragrance and will not create a pleasing aroma before the Lord. Apart from the types I've mentioned, do not use any other kind, as they do not produce a sweet scent.

Follow these instructions carefully, my son, so that you will do what is right in all things. Keep yourself clean at all times. Wash with water before approaching the altar. Wash your hands and feet before and after offering a sacrifice. Make sure no blood remains on your body or clothes. Be careful when handling blood—always cover it with dust. Never eat blood, for it holds the life of the creature. Do not consume any blood at all.

Never accept money or bribes in exchange for a person's life, so that innocent blood is not shed without justice. When blood is spilled, the land becomes polluted, and it can only be cleansed by the blood of the one responsible. Do not take payment to excuse the taking of a life. A life must be paid for with life so that you remain right before the Lord, the Most High God, who protects those who do good. May He shield you from evil and save you from all harm.

My son, I have seen that people's actions are full of sin and wickedness. Their ways are unclean, filled with evil and corruption. There is no righteousness among them. Be careful not to follow their ways or imitate their behavior. Do not commit sins that lead to death before the Most High God. If you do, He will turn away from you, allow you to fall into your own wrongdoing, and remove you and your descendants from the land. Your name and family line will be erased from the earth.

Stay far from their sinful ways and unclean actions. Follow the laws of the Most High God, obey His will, and live righteously in all things. If you do this, He will bless your work and raise a righteous family from you for generations to come. Your name and mine will never be forgotten under heaven and will last forever.

Go in peace, my son. May the Most High God—my God and your God—give you strength to do His will. May He bless your descendants and all future generations with righteousness so that you and your family will be a blessing to the whole earth."

With that, Abraham departed, his heart full of joy.

Chapter XXII.

In the second year of the first week of the forty-fourth jubilee, the same year Abraham passed away, Isaac and Ishmael traveled from the Well of the Oath to celebrate the Feast of Weeks—the festival of the first fruits of the harvest—with their father, Abraham. Seeing both of his sons together made Abraham very happy. Isaac, who owned a lot of land in Beersheba, often visited his property before returning to his father. Around that time, Ishmael also came to visit, and the two brothers reunited.

Isaac offered a burnt sacrifice on the altar that Abraham had built in Hebron. He also presented a thank offering and shared a joyful feast with his brother Ishmael. Rebecca baked fresh cakes from the new grain and gave them to Jacob to bring to Abraham so he could eat and bless the Creator before his passing. Isaac also sent a generous thank offering with Jacob for Abraham to enjoy. Abraham ate and drank, then praised the Most High God, saying:

"Blessed is the Creator of heaven and earth,
Who made all the good things of this world,
And gave them to people,
So they may eat, drink, and give thanks to their Creator."

Then he continued, "I thank You, my God, for letting me see this day. I am now 175 years old, and my life has been full. I have lived in peace, and no enemy has been able to harm me in all that You have given to me and my children. My God, may Your kindness and peace be with me and my descendants. May they be a chosen people, set apart as Yours among all the nations for generations to come."

Then Abraham called Jacob and said, "My son Jacob, may the God who created everything bless you and give you the strength to live righteously and follow His ways. May He choose you and your descendants to be His people forever. Come closer, my son, and give me a kiss."

Jacob stepped forward and kissed him, and Abraham said:
"Blessed be my son Jacob,
And blessed be all the children of the Most High God forever.
May God give you righteous descendants,
And may He choose some of your children to be holy among
    the nations.
May other nations serve you,
And may all people respect your family.
Be strong in the presence of others,
And lead the descendants of Seth.
Through you, righteousness will continue,
And you will become a holy nation."

"May the Most High God bless you with all the blessings
He gave to me, to Noah, and to Adam.

May these blessings remain with your descendants forever.
May He cleanse you from all sin and impurity,
Forgiving any wrongs you have done without knowing.
May He strengthen and bless you.
May you inherit the whole earth,
And may He renew His covenant with you,
Making you His chosen people forever.
May He always be your God,
And the God of your descendants,
In truth and righteousness, for all time."

"Remember my words, my son Jacob,
And always follow the commandments of your father, Abraham.
Do not mix with the other nations.
Do not eat with them or follow their customs.
Do not form close friendships with them,
For their ways are corrupt and sinful.
They worship the dead and follow evil spirits.
They even eat meals near graves,
And their actions are meaningless.
They do not understand,
And they say to a piece of wood, 'You are my god,'
And to a stone, 'You are my lord and savior.'
They are blind to the truth of their own actions."

"My son Jacob, may the Most High God guide and bless you.
May He keep you away from their wickedness and sinful ways.
Do not marry any of the daughters of Canaan,
For their descendants are doomed to destruction.
Because of Ham's sin, the line of Canaan will be completely
    wiped out.
None of them will be saved on the Day of Judgment.

Idol worshipers and those who live in impurity
Will have no place in the land of the living.
They will be forgotten on earth
And sent to Sheol, the place of judgment,
Just like the people of Sodom, who were completely destroyed."

"Do not be afraid, my son Jacob.
Be strong, my child, a descendant of Abraham.
May the Most High God protect you from harm
And rescue you from evil.
This house I have built carries my name on the earth,
And it belongs to you and your descendants forever.
It will be known as the house of Abraham.
You will honor my name before God forever,
And your descendants will carry my name
Through all generations of the earth."

After Abraham finished blessing Jacob and giving him instructions, they lay together on one bed. Jacob rested in Abraham's arms, and Abraham kissed him seven times, filled with love and joy for his grandson. With all his heart, he blessed Jacob, saying:

"May the Most High God, the Creator of all things,
Who brought me out of Ur of the Chaldees to inherit this land
    forever,
Bless my holy descendants. Blessed be the Most High forever."

Then he said to Jacob, "My dear son, who brings me great joy, may God's kindness and grace always be with you and your children. May He never leave you or turn away from you. May His eyes always watch over you and your family. May He protect and bless you, choosing you as His own people. May He give you every lasting blessing, renewing His promise with you and your descendants for generations to come, according to His perfect plan."

Chapter XXIII.

Jacob rested in Abraham's arms, unaware that his grandfather had passed away. Abraham gently placed two of Jacob's fingers over his eyes, blessed the God of all gods, covered his own face, stretched out his feet, and peacefully passed away, joining his ancestors.

When Jacob woke up, he noticed that Abraham's body was cold, like ice. Alarmed, he cried out, "Grandfather, Grandfather!" But there was no response. Realizing that Abraham had died, Jacob quickly ran to his mother, Rebecca, and told her what had happened. Rebecca then went to Isaac in the night and informed him. Together with Jacob, who carried a lamp, they entered the room and found Abraham lying still, lifeless.

Isaac fell over his father's body, weeping with deep sorrow, and kissed him. Soon, the entire household was filled with the sound of mourning. Ishmael, Abraham's son, also arrived and wept for his father. Everyone in the house grieved together, crying loudly from the depths of their hearts.

Later, Isaac and Ishmael buried Abraham in the cave near his wife, Sarah. For forty days, the men of the household mourned him. This included Isaac, Ishmael, their sons, and the sons of Keturah, each grieving in their own places. When the mourning period ended, it was recorded that Abraham had lived a total of 175 years—three jubilees and four weeks of years. He had lived a full life and passed away peacefully, satisfied with his years.

In the past, people had lived much longer—up to nineteen jubilees—but after the Flood, lifespans began to shorten. No one lived that long anymore, as people aged faster, suffered more hardships, and faced the increasing wickedness of the world. Abraham was different, living a life pleasing to God, blameless in his actions. Yet, even he did not reach four full jubilees because of the growing evil on earth, which shortened his days.

From that time forward, human lifespans would continue to decrease. By the time of the final judgment, people would no longer live for even two full jubilees. As they aged, their knowledge would fade, and their understanding would weaken. A man who lived a jubilee and a half would be considered old, yet most of his life would be filled with hardship, sorrow, and suffering, with little peace.

Disaster would come one after another—wounds upon wounds, trouble upon trouble, and endless bad news. People would suffer from sickness, famine, exhaustion, war, captivity, and countless other struggles. These misfortunes would fall upon a generation filled with wickedness, a generation whose actions were sinful and corrupt.

In those days, people would complain, saying, "Our ancestors lived long and good lives—up to a thousand years. But now, we only live seventy or eighty years if we are strong, and our days are full of suffering. There is no peace in this evil generation."

Children would turn against their parents and elders, blaming them for their troubles. They would abandon the covenant that the Lord had made with their ancestors, refusing to follow His commandments or walk in His ways. They would completely turn away from God, caring only about themselves.

Everyone would chase after evil, and every word spoken would be full of lies. Their actions would be corrupt and disgusting, leading only to destruction. The earth would suffer because of their wickedness—vineyards would dry up, oil would disappear, and their unfaithfulness would bring ruin. Humanity, along with animals, livestock, birds, and sea creatures, would suffer because of the sins of mankind.

People would rise up against each other—young against old, the poor against the rich, the humble against the powerful, and beggars against rulers—all because they had abandoned the law and the covenant. They would forget the sacred commandments, the holy festivals, the Sabbaths, the jubilees, and all of the Lord's instructions.

Armed with weapons, they would fight in the hope of finding righteousness again, but they would never return to the right path until the earth was covered in blood. One person would turn against another in endless violence. Even those who survived would refuse to change. Instead, they would be filled with pride and greed, seeking only to take from others. They would claim the name of the Lord but live without truth or righteousness, corrupting even the most sacred places with their sin.

A great punishment would fall upon this generation. The Lord would allow them to fall into war, captivity, and suffering. He would bring ruthless foreign nations against them—people without mercy, who would not spare the old or the young. These invaders would be more wicked and powerful than any before them, bringing destruction to Israel and committing terrible sins against Jacob's descendants.

The land would be covered in blood, and there would be no one left to bury the dead.

In those days, people would cry out for help, calling for rescue from sinners and oppressors, but no one would come to save them.

The heads of children would turn gray with age,
And even babies as young as three weeks old would appear as
    old as men of a hundred years,
Worn down by suffering and endless hardship.
Yet in those days, some would begin to seek the Lord again.
Children would study the law,

Search for the commandments,
And return to the path of righteousness.

Lifespans would grow longer,
And people would once again live for many years,
Almost reaching a thousand years,
Just as in the days of old.

No one would be considered old,
And no one would feel like their life was too short.
Everyone would have the strength of youth and the joy of
    childhood.

Their days would be filled with peace and happiness,
For Satan and all evil forces would be removed from the earth.
Blessings and healing would fill their lives.
In that time, the Lord would restore health to His people,
And they would rise up in joy,
Driving away their enemies.

The righteous would celebrate with gratitude,
Lifting their voices in endless praise to the Lord,
Witnessing His justice and the defeat of their enemies.

Their bodies would rest in the earth,
But their spirits would rejoice forever,
Knowing that the Lord is a just and merciful judge,
Showing kindness to all generations who love Him.
Then the Lord said to Moses, "Write down these words, for
    they are recorded on the heavenly tablets as a message for
    future generations."

Chapter XXIV.

In the first year of the third week of the forty-fourth jubilee, after Abraham passed away, the Lord blessed his son Isaac. Isaac left Hebron and moved to live near the Well of the Vision, where he stayed for seven years. In the first year of the fourth week, a famine struck the land, just like the one that had happened during Abraham's time.

One day, Jacob was cooking a pot of lentils when Esau came in from the fields, exhausted and starving. Esau said, "Give me some of that red stew." Jacob replied, "Sell me your birthright, and I will give you bread and stew." Esau thought, "I'm so hungry I could die— what good is my birthright to me?" So he agreed, swearing an oath and giving his birthright to Jacob. Jacob then gave him bread and stew, which Esau ate until he was full. Esau did not care about his birthright, and because of this, he was called Edom, named after the red stew for which he had traded his inheritance. From that moment on, Jacob became the rightful firstborn, and Esau lost his position.

When the famine ended, in the second year of that week, Isaac planned to go to Egypt but instead went to Gerar, where Abimelech, the king of the Philistines, ruled. The Lord appeared to Isaac and said, "Do not go to Egypt. Stay in the land I show you and live here. I will be with you and bless you. I will give this land to you and your descendants and fulfill the promise I made to Abraham, your father. I will make your family as numerous as the stars in the sky and give them all this land. Through your descendants, all nations will be blessed because Abraham listened to My voice, obeyed My commandments, and followed My laws. Now, you must do the same and remain here."

Isaac stayed in Gerar for 21 years. While he was there, Abimelech warned his people, "Anyone who harms Isaac or takes anything that belongs to him will be put to death." Isaac became successful among the Philistines, gaining many possessions, including oxen, sheep, camels, donkeys, and many servants. He planted crops in the Philistine land and harvested a hundred times more than expected, becoming very wealthy. This made the Philistines jealous.

Out of envy, the Philistines filled the wells that Abraham's servants had dug with dirt. Then Abimelech said to Isaac, "Leave us because you have become too powerful for us." So in the first year of the seventh week, Isaac moved to the valleys of Gerar. There, his servants reopened the wells that Abraham had dug, which the Philistines had filled after his death. Isaac named the wells just as his father had.

Isaac's servants dug a new well in the valley and found fresh water. But the shepherds of Gerar argued with them, saying, "This water belongs to us." So Isaac named the well "Dispute" because of their unfair claim. His servants dug another well, but the locals fought over it too, so he named it "Opposition."

Isaac then moved on and dug another well, and this time, no one fought over it. He named it "Room" and said, "The Lord has made space for us, and we have prospered in this land."

In the first year of the first week of the forty-fourth jubilee, Isaac moved to the Well of the Oath. That night, on the new moon of the first month, the Lord appeared to him and said, "I am the God of Abraham, your father. Do not be afraid, for I am with you. I will bless you and make your descendants as numerous as the sand of the earth because of My servant Abraham." Isaac built an altar at the same place where his father had built one. He called on the name of the Lord and offered a sacrifice. His servants then dug another well and found fresh water.

Later, Isaac's servants dug another well but found no water. They told Isaac, and he said, "I have sworn peace with the Philistines, and they know about this well." He named the place "Well of the Oath" because of the peace agreement he had made with Abimelech, Ahuzzath, and Phicol, the commander of their army. Isaac understood that he had no choice but to make peace with them.

On that day, Isaac cursed the Philistines, saying:

"The Philistines will be cursed until the day of judgment when they are scattered among the nations. They will be a disgrace and a curse, hated and punished by sinful nations and the Kittim. Even if they escape the enemy's sword and the Kittim, they will still face judgment from the righteous nation. They will remain enemies of my descendants forever. When the day of judgment arrives, their entire bloodline will be wiped out, and no one from the Caphtorim will be left on earth.

If they try to rise to power, they will be brought down. If they become strong, they will be uprooted. If they hide among other nations, they will be found and removed. If they descend into Sheol, they will suffer greatly and never find peace. If they are taken as captives, those who pursue them will destroy them before they can escape. No descendants will remain, and their name will be erased from history. They will be cursed forever."

This decree is written on the heavenly tablets and will be fulfilled on the day of judgment, when the Philistines will be completely wiped out from the earth.

Chapter XXV.

In the second year of that week, during this jubilee, Rebecca called her son Jacob and said, "My son, do not marry one of the daughters of Canaan like your brother Esau. He has married two women from this land, and they have brought me nothing but sorrow. Their ways are immoral and sinful, and there is no goodness in them. Everything they do is filled with wickedness, and they bring only grief and pain.

I love you deeply, my son, and I bless you every moment, day and night. Please listen to me and follow my advice. Do not choose a wife from the women of this land. Instead, find a wife from my father's family, from among our own people. If you marry within our family, the Most High God will bless you, and your children will be righteous and holy."

Jacob answered his mother, "Mother, I am still young, only nine weeks old, and I have no knowledge or experience with women. I have not made any promises to anyone, and I do not plan to marry a daughter of Canaan.

I remember what our father Abraham told me. He warned me not to marry a woman from Canaan but to choose a wife from our own family and people. I know that your brother Laban has daughters, and I would like to marry one of them.

I have kept myself pure for this reason, staying away from sin and corruption. Father Abraham gave me clear instructions to avoid lust and wrongdoing, and I have followed his guidance.

Even though my brother Esau has pressured me many times to marry one of his wives' sisters, I have refused to follow his example. I promise you, mother, I will never marry a woman from Canaan, and I will not act wickedly as my brother has done.

Do not worry, mother. I will honor your wishes and live a righteous life, never straying from the right path."

Hearing this, Rebecca lifted her eyes to heaven, stretched out her hands, and prayed, thanking and praising the Most High God, the Creator of heaven and earth. She said,

"Blessed be the Lord God, and may His holy name be praised forever. He has given me Jacob, a pure son, a righteous seed. He belongs to You, Lord, and his descendants will be Yours for all generations to come.

Bless him, O Lord, and let my words carry the blessing of righteousness as I speak over him."

Then, filled with the spirit of righteousness, she placed her hands on Jacob's head and said,

"Blessed are You, Lord of righteousness and God of all ages. May You bless Jacob above all the people of the earth.

Guide him, my son, on the path of righteousness, and may his descendants also walk in truth and goodness.

May his children multiply like the months of the year and become as numerous as the stars in the sky, outnumbering the grains of sand by the sea.

Give them this good land, as You promised to Abraham and his descendants, and may they possess it forever.

May I live to see your children, my son, and may all your descendants be holy and blessed.

Just as you have brought me joy and comfort, may you be blessed by the womb that bore you, by the love of my heart, by the milk that nourished you, and by the words of my mouth that praise you always.

May you grow and spread across the earth, and may your descendants rejoice forever in heaven and on earth.

May your children be perfect, full of joy, and find peace on the great day of peace.

May your name and your family line last forever. May the Most High God always be your God, and may the God of righteousness be with your descendants. Through them, may His holy place be established for all generations.

Blessed are those who bless you, and cursed are those who falsely curse you."

Rebecca kissed Jacob and said, "May the Lord of all creation love you as much as my heart loves you. May my joy and blessings always remain with you." Then, she finished her blessings.

Chapter XXVI.

In the seventh year of that week, Isaac called his older son, Esau, and said, "My son, I am old, and my eyesight is failing. I do not know how much longer I will live. Take your bow and arrows, go to the fields, and hunt some wild game for me. Prepare my favorite meal and bring it to me so that I may eat and bless you before I die."

Rebecca overheard Isaac speaking to Esau. After Esau left early to hunt, she called Jacob and said, "I heard your father tell Esau, 'Go hunt for me and prepare my favorite meal so I can eat and bless you before the Lord before I die.' Now listen carefully, my son. Go to the flock and bring me two young goats. I will prepare them just the way your father likes. Then you will take the food to him, and he will eat and bless you before he dies. This way, you will receive the blessing instead."

Jacob hesitated and said, "But, Mother, what if my father recognizes my voice or touches me? I have smooth skin, and Esau is hairy. If he touches me, he will know I am tricking him, and instead of a blessing, I will bring a curse upon myself."

Rebecca replied, "Let any curse fall on me, my son. Just do what I tell you."

So Jacob obeyed and brought the two young goats. Rebecca prepared the food exactly how Isaac liked it. Then she took Esau's best clothes, which she had in the house, and dressed Jacob in them. She covered his hands and the smooth part of his neck with goat skins so that he would feel hairy like Esau. Then she handed Jacob the food and bread she had made.

Jacob went to his father and said, "Father, I have done as you asked. Please sit up and eat the meal I prepared so that you may bless me."

Isaac asked, "How did you find the game so quickly, my son?"

Jacob answered, "Because the Lord your God helped me succeed."

Isaac said, "Come closer so I can touch you, my son, and see if you really are Esau."

Jacob moved closer, and Isaac touched him. He said, "The voice is Jacob's, but the hands feel like Esau's." Isaac did not recognize him because his hands felt hairy like Esau's. It was by God's will that Isaac could not tell the difference, so he continued to bless him.

"Are you really my son Esau?" Isaac asked.

Jacob answered, "Yes, I am."

Isaac said, "Bring me the food so I may eat and bless you."

Jacob brought him the meal and wine, and Isaac ate and drank. Then Isaac said, "Come closer, my son, and kiss me."

When Jacob leaned in to kiss him, Isaac smelled the scent of Esau's clothes and blessed him, saying:

"My son smells like a field that the Lord has blessed.
May God give you rain from the sky,
Rich soil,
And plenty of grain and wine.

May nations serve you,

And people show you honor.

Be a ruler over your brothers,

And may your mother's sons respect you.

Anyone who curses you will be cursed,

And anyone who blesses you will be blessed."

Just as Jacob left, Esau returned from his hunt. He prepared the meal and brought it to his father, saying, "Father, sit up and eat the venison I've brought so you can bless me."

Isaac asked, "Who are you?"

Esau replied, "I am your firstborn son, Esau."

Isaac trembled and said, "Then who was it that hunted game and brought it to me? I already ate it, and I have blessed him—and the blessing will stand!"

When Esau heard this, he let out a loud and bitter cry. "Father, bless me too!" he pleaded.

Isaac said, "Your brother came with deceit and took your blessing."

Esau cried out, "No wonder his name is Jacob! He has cheated me twice—first, he took my birthright, and now he has taken my blessing! Haven't you saved any blessing for me?"

Isaac replied, "I have made him lord over you and given him all his brothers as servants, along with plenty of grain and wine. What more can I give you, my son?"

Esau begged, "Father, do you only have one blessing? Bless me too!" And he wept loudly.

Isaac said,

"You will live far from the fertile land
And without the blessing of rain from above.
You will survive by the sword
And serve your brother.

But when you can no longer bear it,
You will break free from his control.
However, your choices will lead to great sin,
And your descendants will be lost."

Esau was filled with hatred toward Jacob because of the blessing his father had given him. He said to himself, "Soon my father will die, and then I will kill my brother Jacob."

Chapter XXVII.

Rebecca had a dream warning her about Esau's plan to take revenge and kill Jacob. She immediately called Jacob and said, "Your brother Esau is planning to harm you. Listen to me, my son, and do what I say. Leave right away and go to my brother Laban in Haran. Stay there for a while until Esau calms down and forgets what happened. Then I'll send for you to return."

Jacob replied, "I'm not afraid of him. If he comes after me, I'll defend myself and fight back."

But Rebecca said, "I don't want to lose both of you in one day."

Jacob answered, "You know that Father is old and nearly blind. If I leave without his blessing, it will upset him, and he might curse me instead of bless me. I won't go unless he sends me himself."

Rebecca reassured him, "I will talk to him, and he will bless you before you leave."

She went to Isaac and said, "I am deeply troubled by Esau's wives.

If Jacob marries a woman like them, I don't want to live anymore. The women of this land are full of wickedness."

Isaac then called Jacob, blessed him, and said, "Do not marry a woman from Canaan. Go to Mesopotamia, to your mother's family, and find a wife from the daughters of your uncle Laban. May God Almighty bless you, give you many children, and make you into a great nation. May He grant you the blessings promised to Abraham, and may you inherit this land, where you now live as a foreigner— the land God gave to Abraham. Go in peace, my son."

Isaac sent Jacob on his way, and Jacob traveled to Mesopotamia, to the house of Laban, the son of Bethuel, Rebecca's brother.

After Jacob left, Rebecca was filled with sorrow and wept for her son.

Isaac comforted her, saying, "Do not cry for Jacob, my love. He left in peace, and he will return in peace. The Most High God will watch over him, protect him, and guide him. He will succeed in all he does, and when he comes back, we will rejoice with him again. Be at ease, for Jacob is righteous and follows the path of truth. He will not perish. Wipe your tears."

Isaac's words reassured Rebecca, and they both blessed Jacob.

Jacob left the well of Beer-sheba and began his journey toward Haran in the first year of the second week of the 44th jubilee. He arrived at a place called Luz, later known as Bethel, on the new moon of the first month. As the sun was setting, he turned off the road and decided to spend the night there.

He took a stone from the area, placed it under his head as a pillow, and lay down to sleep.

That night, Jacob had a dream. He saw a ladder reaching from the ground up to heaven, with angels going up and down on it. At the top of the ladder stood the Lord, who spoke to him:

"I am the Lord, the God of your grandfather Abraham and the God of your father Isaac. The land where you are lying, I will give to you and your descendants. Your family will be as countless as the dust of the earth, spreading in all directions—west, east, north, and south. Through you and your descendants, all the families of the earth will be blessed.

I am with you and will protect you wherever you go. I will bring you back to this land, and I will not leave you until I have done everything I have promised."

When Jacob woke up, he said, "Surely, the Lord is in this place, and I didn't even realize it." Filled with awe, he added, "This place is amazing. It is truly the house of God and the gateway to heaven."

Early the next morning, Jacob took the stone he had used as a pillow, stood it upright as a pillar, and poured oil over it. He named the place Bethel, though it was originally called Luz.

Then Jacob made a vow, saying, "If God will be with me, protect me on this journey, provide food and clothing, and bring me back safely to my father's house, then the Lord will be my God. This stone I have set up will be the house of God, and I will give a tenth of everything You bless me with."

Chapter XXVIII.

Jacob continued his journey and arrived in the land of the east, at the home of Laban, his mother Rebecca's brother. He stayed there and worked for seven years so he could marry Rachel, Laban's daughter. After completing the seven years, Jacob said to Laban, "I have worked as promised. Now give me Rachel as my wife."

Laban agreed and prepared a wedding feast, but instead of giving Rachel to Jacob, he gave him his older daughter, Leah. He also gave Leah a servant named Zilpah to be her maid. Jacob, unaware of the trick, spent the night with Leah, believing she was Rachel. When he realized the truth, he was furious and said to Laban, "Why have you deceived me? I worked for Rachel, not Leah. This is wrong! Take Leah back and let me go."

Jacob loved Rachel more than Leah. Leah had soft eyes, but Rachel was very beautiful and had a graceful figure. Laban told Jacob, "It is not our custom to marry off the younger daughter before the older one. This rule is also written in the heavenly records—it is a sin to break it." He warned Jacob never to go against this law in the future.

Laban then said, "Finish the wedding celebrations for Leah, and after one week, I will also give you Rachel. But in return, you must work for me another seven years."

Jacob agreed, and after the seven-day celebration for Leah, Laban gave Rachel to him as well. He also gave Rachel a servant named Bilhah to be her maid. Jacob then worked another seven years to fulfill his promise for Rachel.

The Lord saw that Leah was unloved, so He blessed her with children. In the first year of the third week, Leah gave birth to a son and named him Reuben. But Rachel remained childless because the Lord had not yet given her children, as Jacob loved her more than Leah.

Leah became pregnant again and had a second son, Simeon. Later, she had a third son and named him Levi. Then she gave birth to a fourth son and called him Judah.

Meanwhile, Rachel became jealous of Leah for having children and said to Jacob, "Give me children, or I will die!"

Jacob answered, "I am not God. He alone decides who can have children."

Seeing that Leah had four sons—Reuben, Simeon, Levi, and Judah—Rachel said to Jacob, "Take my servant Bilhah as a wife. She will have children for me."

Rachel gave Bilhah to Jacob, and Bilhah became pregnant and had a son. Rachel named him Dan. Bilhah became pregnant again and had another son, whom Rachel called Naphtali.

Leah noticed that she had stopped having children, so she gave her servant Zilpah to Jacob as a wife. Zilpah gave birth to a son, and Leah named him Gad. Then Zilpah had another son, and Leah named him Asher.

Later, Leah became pregnant again and gave birth to a son named Issachar. She had another son, whom she called Zebulun, and then a daughter, Dinah.

Finally, the Lord answered Rachel's prayers and allowed her to have a child. She became pregnant and gave birth to a son, naming him Joseph.

After Joseph was born, Jacob said to Laban, "Let me take my wives and children and return to my father's house. I have served you faithfully, and now I wish to go home."

Laban replied, "Stay with me a little longer. Name your wages, and I will pay you. Continue caring for my flocks."

They agreed that Jacob's payment would be all the black, spotted, and speckled lambs and goats born among Laban's flocks. Over time, more and more animals were born with these markings, increasing Jacob's share.

Jacob's wealth grew, and he gained many sheep, oxen, camels, donkeys, servants, and maids. However, Laban and his sons became jealous of Jacob's success. Laban started taking back some of the sheep and treated Jacob unfairly.

Chapter XXIX.

After Rachel gave birth to Joseph, Laban left to shear his sheep, which were three days away from his home. Seeing this as the perfect time to leave, Jacob called Leah and Rachel and spoke to them kindly, asking them to go with him back to the land of Canaan. He told them about a dream in which God instructed him to return to his father's house. Leah and Rachel agreed, saying, "We will go wherever you go."

Jacob then praised the God of his father, Isaac, and his grandfather, Abraham. Early in the morning, he gathered his wives, children, and all his belongings and crossed the river. Without telling Laban, they set out toward Gilead.

In the seventh year of the fourth week, on the twenty-first day of the first month, Jacob began his journey to Gilead. When Laban realized Jacob had left, he chased after him and caught up in the mountains of Gilead on the thirteenth day of the third month. However, God appeared to Laban in a dream that night and warned him not to harm Jacob.

On the fifteenth day, Jacob prepared a feast for Laban and his men. During the gathering, Jacob and Laban made an agreement, promising never to cross the mountains of Gilead with the intent to harm one another. To mark their promise, Jacob built a mound of stones as a witness to their covenant. The place was named "The Heap of Witness."

Before this, the region of Gilead had been home to the Rephaim, a race of giants who were between seven and ten cubits tall. Their land stretched from the territory of the Ammonites to Mount Hermon, including cities like Karnaim, Ashtaroth, Edrei, Misur, and

Beon. Because of their wickedness, God wiped them out, and the Amorites later took over their land. No people since have sinned to the same extent as the Rephaim, and their time on earth was cut short.

After making peace, Jacob sent Laban on his way, and Laban returned to Mesopotamia. Jacob continued his journey and crossed the Jabbok River on the eleventh day of the ninth month. That same day, his brother Esau came to meet him. The two brothers reconciled and made peace. Esau then traveled back to the land of Seir, while Jacob remained living in tents.

In the first year of the fifth week, Jacob crossed the Jordan River and settled on the other side. He grazed his flocks from the Sea of the Heap to Bethshan, Dothan, and the Akrabbim forest. From his wealth, Jacob sent gifts to his father Isaac, including clothes, food, meat, drink, milk, butter, cheese, and dates from the valley.

He also sent gifts to his mother Rebecca four times a year—after plowing, during harvest, after autumn, and in the spring. He delivered them to the tower of Abraham, where Rebecca lived.

Meanwhile, Isaac had moved from the Well of the Oath to the tower of Abraham in the mountains of Hebron, living separately from Esau. During Jacob's time in Mesopotamia, Esau married Mahalath, Ishmael's daughter. Later, Esau moved to Mount Seir with his flocks and wives, leaving Isaac behind at the Well of the Oath. Isaac then relocated to the tower of his father, Abraham.

Jacob continued to provide for his father and mother, sending them whatever they needed. In return, Isaac and Rebecca blessed him with all their hearts and souls.

Chapter XXX.

In the first year of the sixth week, during the fourth month, Jacob settled in Salem, east of Shechem, arriving safely. While there, Shechem, the son of Hamor the Hivite, the ruler of the land, took Jacob's young daughter Dinah into his house and violated her. She was only twelve years old. Afterward, Shechem wanted to marry Dinah, so he approached his father and then Jacob and his sons with a proposal.

Jacob and his sons were furious about what had been done to Dinah. Although they pretended to respond peacefully, they secretly planned revenge.

Simeon and Levi launched a surprise attack on Shechem, killing all the men in the city to avenge their sister. They left no one alive, making it clear that no daughter of Israel should ever be treated this way again. It was declared in heaven that anyone who committed such a crime deserved to die, for their actions brought disgrace to Israel.

The Lord allowed Jacob's sons to carry out this judgment to prevent others from committing the same sin in the future. It was also commanded that if any Israelite man gave his daughter or sister to a foreigner, he would be put to death by stoning. Likewise, any Israelite woman who dishonored her family by marrying outside of Israel would be burned and removed from the nation.

Israel was called to remain pure before the Lord, avoiding all impurity and unfaithfulness. Anyone who corrupted the nation would face death, as recorded in the heavenly laws. No forgiveness or atonement would be granted to those who violated these laws. Anyone who allowed impurity to spread or gave their children to foreign customs was committing a serious sin and would be cut off from the nation.

Moses was instructed to warn the people of Israel never to marry Gentiles or allow their daughters to do so, for it was considered a terrible offense before the Lord. The story of Shechem was recorded as a warning, showing the judgment carried out by Jacob's sons, who declared, "We will never give our sister to an uncircumcised man, for it would be a disgrace to us."

Marrying outside of Israel was seen as sinful and shameful. Such actions would bring plagues and curses upon the nation. Anyone who ignored this impurity or allowed it to happen would also face judgment and punishment. The entire community would suffer because of it, and the Lord would reject any offerings, sacrifices, or incense from those who committed these sins.

This is why Moses was commanded to teach Israel about the Shechemites and how they were judged by Jacob's sons. Their actions were seen as righteous, and they were written in the heavenly records as a blessing. Levi, in particular, was honored for his dedication to justice. Because of this, he and his descendants were chosen to serve as priests for the Lord forever. His name was recorded in the heavenly tablets as a righteous man and a faithful servant of God.

These events were recorded to remind Israel of their covenant and the importance of following God's laws. If they obeyed, they would be counted as friends of the Lord. But if they sinned and broke the covenant, they would be seen as enemies, erased from the book of life, and placed among those destined for destruction.

On the day Jacob's sons destroyed Shechem, their actions were recorded in heaven as righteous judgment. They rescued Dinah from Shechem's house and took everything from the city—the livestock, goods, and wealth—including sheep, oxen, and donkeys—bringing it all back to Jacob.

Although Jacob was upset with his sons for attacking the city, fearing that the neighboring Canaanites and Perizzites would seek revenge, the fear of the Lord fell upon those nearby. No one dared to attack Jacob's family because terror had spread among the surrounding cities.

Chapter XXXI.

At the beginning of the month, Jacob gathered his family and said, "Purify yourselves and change into clean clothes. We are going to Bethel, the place where I made a vow to God when I fled from my brother Esau. God has been with me, protected me, and brought me safely back to this land. Now, get rid of any foreign gods among you."

His family handed over their idols, including the ones Rachel had taken from her father Laban, as well as the jewelry they wore in their ears. Jacob collected everything, destroyed them, and buried the remains under an oak tree in Shechem.

Jacob then traveled to Bethel at the beginning of the seventh month. He built an altar at the place where he had once slept and set up a pillar in honor of the Lord. He sent a message to his father, Isaac, inviting him and his mother, Rebekah, to join him for the sacrifice. Isaac replied, "Let my son Jacob come to me so I may see him before I die."

Jacob brought his sons, Levi and Judah, to visit Isaac. When Rebekah heard that Jacob had arrived, she left the tower and ran to meet him. Her heart filled with joy when she heard, "Jacob, your son, has returned." She embraced him tightly, kissed him, and wept with happiness. Then she saw Levi and Judah and asked, "Are these your sons, my child?" She hugged and kissed them, blessing them, and said, "Through you, the descendants of Abraham will become a great people, bringing blessings to the world."

Jacob entered Isaac's room, where his father lay resting. He took

110

Isaac's hand, leaned down, and kissed him. Isaac held Jacob close and wept. Though his eyesight was failing, his spirit lifted, and he asked, "Are these your sons? They look just like you." Jacob confirmed that they were.

Isaac kissed both boys, and suddenly, the spirit of prophecy filled him. He took Levi's right hand and Judah's left, then began to bless Levi first.

"May the eternal God bless you and your children. May you be chosen and set apart to serve in His holy place, just like the angels in heaven. Your family will be honored and respected, always serving in God's presence. Your descendants will lead and guide the tribes of Israel, teaching them His laws and commandments. His blessings will be spoken through you, and you will bring goodness to all His people."

Isaac continued, "Your mother named you Levi, and your name is fitting because you will always be connected to the Lord. You will share in the inheritance with Jacob's sons, and your family will receive blessings from God's table. May you never be in need and always have more than enough. Anyone who stands against you will fail, and those who curse you will be cursed, but those who bless you will be blessed."

Then Isaac turned to Judah and said, "May the Lord give you the strength to defeat your enemies. You will be a leader among your brothers, and from your family will come a ruler who will govern the descendants of Jacob. Your name will be known everywhere, and nations will respect your family. Through you, Jacob's people will find help, and Israel will be saved. When you rule with fairness, peace will spread to all of God's people. Those who bless you will be blessed, and those who stand against you will fade away."

Isaac kissed Judah, hugged him, and felt great joy at seeing Jacob's sons. He blessed them once more and then rested at Jacob's feet.

That evening, Isaac and Jacob ate together, celebrating with happiness. Jacob placed Levi and Judah on either side of Isaac as they slept, and Isaac felt it was a righteous act.

That night, Jacob shared stories with Isaac about how the Lord had been with him, protected him, and blessed him. Isaac praised the God of Abraham for showing mercy to Jacob and his family.

The next morning, Jacob told Isaac about the vow he had made at Bethel, describing the vision he had seen and the altar he had built. Isaac said, "I am too old to travel, my son. I am 165 years old and no longer strong enough to make the journey. Take your mother with you and fulfill the vow you made to the Lord without delay. Be faithful to it, for you are responsible for keeping your promise. May the Creator of all things accept your offering and be pleased."

Isaac instructed Rebekah to go with Jacob, and she traveled with him, bringing her servant Deborah. As they made their way to Bethel, Jacob reflected on Isaac's blessings over Levi and Judah, and his heart was filled with gratitude. He praised the God of his fathers, Abraham and Isaac, saying, "Now I know that my future is secure, and that my sons will also be blessed before the Lord forever."

This moment was recorded in the heavenly books as an eternal testimony of Isaac's blessings over Levi and Judah.

Chapter XXXII.

That night, Jacob stayed in Bethel, and Levi had a dream where God chose him and his descendants to serve as priests for the Most High forever. When he woke up, Levi praised and thanked God for this great honor.

The next morning, on the fourteenth day of the month, Jacob dedicated a tenth of everything he owned—his servants, livestock, gold, silver, and all his possessions—as an offering to God. Around this time, Rachel became pregnant with her son, Benjamin. Jacob counted his sons and decided that Levi's share would be set apart for the Lord. He dressed Levi in priestly robes and gave him the responsibility of serving as a priest.

On the fifteenth day of the month, Jacob made a sacrifice at the altar, offering fourteen oxen, twenty-eight rams, forty-nine sheep, seven lambs, and twenty-one goats, along with grain and drink offerings. This was to fulfill a promise he had made to God. After the fire consumed the sacrifices, Jacob burned incense on the altar as an offering of thanksgiving.

For the next seven days, he continued making offerings, sacrificing two oxen, four rams, four sheep, four goats, and two lambs each day. Jacob, his sons, and his household celebrated with food and drink, thanking God for His kindness and protection.

Jacob also offered a tenth of all his clean animals as a burnt offering to the Lord. However, the unclean animals were not included in Levi's portion. Levi served as a priest at Bethel, leading the offerings before Jacob and his brothers. Jacob renewed his promise to God, dedicating another tithe and setting it apart for the Lord. This became a lasting tradition: every year, the second tithe was to be eaten in the place God chose, with none of it left over. Any tithe not eaten by the end of the year would be considered unclean and burned.

Jacob then planned to build a sacred place at Bethel, enclosing it with a wall to create a permanent sanctuary for himself and his descendants. That night, the Lord appeared to him, blessed him, and said, "Your name will no longer be Jacob but Israel." God promised that Jacob's family would grow into many nations and that kings would come from his descendants. He also reaffirmed that the land under heaven would belong to Jacob's children, who would rule over many nations.

After God finished speaking, Jacob watched as He ascended into heaven. Later that night, Jacob had another dream where an angel came down from heaven carrying seven tablets. The angel handed them to Jacob, and as he read them, he understood everything that would happen to him and his descendants in the future.

The angel told Jacob not to build a permanent sanctuary at Bethel but to return to his father's house until Isaac passed away. The angel also revealed that Jacob would die peacefully in Egypt and be buried with honor alongside Abraham and Isaac.

The angel reassured Jacob not to be afraid, promising that everything he had seen and read would come true. When Jacob worried about remembering all the details, the angel told him that he would recall them when the time was right.

After the vision ended, Jacob woke up, remembering everything he had seen. He wrote it all down and, the next day, made another offering as he had before. He named the day "Addition" because it was added to the yearly feast days, and it became a part of Israel's yearly celebrations.

On the night of the twenty-third of the month, Deborah, Rebekah's nurse, passed away. Jacob buried her near a river by the city, under an oak tree, naming the place "The River of Deborah" and the tree "The Oak of Deborah's Mourning."

Rebekah returned home with Isaac, and Jacob sent gifts of rams, sheep, and goats for them to prepare a meal. Later, Jacob traveled to the land of Kabratan to be near his mother and stayed there for a while.

During this time, Rachel gave birth to a son. Her labor was very painful, and she named him "Son of My Sorrow." However, Jacob changed his name to Benjamin. Sadly, Rachel died during childbirth, and Jacob buried her in Ephrath, which is also called Bethlehem. He set up a pillar on her grave as a marker, and it remained there along the road.

Chapter XXXIII.

Jacob settled with his wife Leah south of Magdaladra'ef on the first day of the tenth month. Around this time, Reuben, Jacob's oldest son, saw Bilhah, Rachel's servant and his father's concubine, bathing in a private place. He was attracted to her and waited until nighttime to act on his desires.

Late at night, Reuben secretly entered Bilhah's tent while she was sleeping alone. He lay with her, and when she woke up and realized what had happened, she was horrified. Bilhah screamed when she recognized Reuben and clung to her blanket in shame. Reuben ran away, leaving her deeply distressed. She was overcome with grief but kept what had happened to herself.

When Jacob returned to Bilhah, she told him everything, saying, "I am no longer pure for you because I have been dishonored. Reuben lay with me while I was asleep, and I didn't know until he uncovered my blanket and defiled me." Jacob was furious with Reuben for committing such a terrible sin, bringing shame upon his father. From that day forward, Jacob never went to Bilhah again.

This act was a great sin in God's eyes. It is strictly forbidden for a man to be with his father's wife or bring shame upon his father in this way. Such behavior is disgraceful and offensive to the Lord. It is written in the heavenly records that no man should ever commit this sin. The punishment for it is death by stoning, and both the man and the woman involved must be removed from the people of God to keep the nation pure.

It is also written, "Cursed is the man who lies with his father's wife, for he has shamed his father." And all the holy ones of the Lord declared, "Amen." Moses was commanded to teach this law to the children of Israel, for it carries the penalty of death. It is a serious impurity, and there is no forgiveness for it. Anyone who commits this sin must be put to death immediately and removed from the community. Such a person must not remain alive even for a single day, as their actions have polluted the people of God.

Although Reuben was not punished while Jacob was alive, it was only because the full law and its judgment had not yet been given. Now, this law is established forever for all generations. There is no atonement for this sin. Anyone who commits it must be removed from the nation and put to death on the same day. Moses was commanded to write down this law so that Israel would follow it and avoid sins that lead to destruction.

The Lord, who is a just Judge, does not show favoritism and cannot be bribed. Moses was to remind the people of this covenant so they would obey it and protect themselves from being cut off from the land. Anyone who commits this sin is considered unclean before God. There is no greater impurity on earth than this kind of wrongdoing. Israel is a holy nation, chosen by God to receive His promises—a people set apart to serve Him. Such impurity must not be found among them.

In the third year of the sixth week, Jacob and all his sons moved to the home of Abraham, near Isaac, his father, and Rebekah, his mother. Jacob's sons were: Reuben, the firstborn; Simeon, Levi, Judah, Issachar, and Zebulun, the sons of Leah; Joseph and Benjamin, the sons of Rachel; Dan and Naphtali, the sons of Bilhah; Gad and Asher, the sons of Zilpah; and Dinah, Leah's only daughter.

When they arrived, they bowed in respect before Isaac and

Rebekah, and Isaac blessed Jacob and his sons. Seeing Jacob's children filled Isaac with joy. He spoke blessings over them, praising God for the family that had grown through Jacob.

Chapter XXXIV.

In the sixth year of the forty-fourth jubilee, Jacob sent his sons and their servants to graze the sheep near Shechem. While they were there, seven kings of the Amorites made a plan to attack them. They hid among the trees, waiting for the right moment to steal their livestock.

At that time, Jacob was at home with Levi, Judah, Joseph, and Isaac because Isaac was feeling sorrowful, and they didn't want to leave him alone. Benjamin, the youngest, also stayed with his father.

The kings of Taphu, Aresa, Seragan, Selo, Ga'as, Bethoron, and Ma'anisakir, who were from Canaan, learned about the Amorites' plot and sent a message to Jacob: "The Amorite kings have surrounded your sons and taken their herds."

Jacob immediately gathered Levi, Judah, Joseph, his father's servants, and his own men, making a total of six thousand warriors armed with swords. They marched to Shechem to face the Amorites. A fierce battle broke out, and Jacob's forces chased down and defeated the Amorites. They killed the kings of Aresa, Taphu, Seragan, Selo, Ga'as, and Ma'anisakir, reclaimed the stolen livestock, and subdued their enemies.

After the battle, Jacob forced the defeated kings to pay tribute by giving five types of fruit from their land. He also built two cities, Robel and Tamnatares, before returning home safely. The defeated kings remained under Jacob's rule until he and his sons later moved to Egypt.

In the seventh year of the next week, Jacob sent Joseph to check on his brothers in Shechem. However, Joseph found them in Dothan, where they had moved their flocks. When Joseph arrived, his brothers plotted against him, intending to kill him. But instead of going through with their plan, they decided to sell him to Ishmaelite merchants. These merchants took Joseph to Egypt and sold him to Potiphar, Pharaoh's chief cook and a priest in the city of 'Elew.

To cover up their actions, Joseph's brothers killed a goat, dipped his coat in its blood, and sent it to Jacob on the tenth day of the seventh month. When Jacob saw the blood-stained coat, he believed that a wild animal had killed Joseph. Overcome with sorrow, he cried, "A wild beast has devoured Joseph!" His entire household mourned with him that day, but Jacob refused to be comforted, saying, "I will grieve for my son until I go to my grave."

That same month, Bilhah, who was living in Qafratef, was so heartbroken after hearing about Joseph's death that she passed away. Soon after, Dinah, Jacob's daughter, also died. Within a single month, Jacob lost Bilhah, Dinah, and Joseph—or so he believed. Both Bilhah and Dinah were buried near Rachel's grave.

Jacob mourned for Joseph for an entire year and could not find peace. Again and again, he repeated, "I will go to my grave mourning for my son."

Later, it was commanded that the people of Israel fast on the tenth day of the seventh month—the day Jacob learned of Joseph's supposed death. This was to be a day of atonement for their sins, marked by the sacrifice of a young goat. It also served as a reminder of the deep sorrow Jacob felt for Joseph and was set as a time for spiritual cleansing.

After Joseph was gone, Jacob's sons began to marry.

- Reuben married Ada.
- Simeon first married Adlba'a, a Canaanite, but later repented and married another wife from Mesopotamia, following the example of his brothers.
- Levi married Melka, a daughter of Aram from Terah's family.
- Judah married Betasu'el, a Canaanite.
- Issachar married Hezaqa.
- Zebulun married Ni'iman.
- Dan married Egla.
- Naphtali married Rasu'u from Mesopotamia.
- Gad married Maka.
- Asher married Ijona.
- Joseph later married Asenath, an Egyptian.
- Benjamin married Ijasaka.

Chapter XXXV.

In the first year of the first week of the forty-fifth jubilee, Rebecca called Jacob to her and gave him advice about his father and brother. She told him to always respect and care for them.

Jacob replied, "I will do everything you ask, Mother. Treating them with respect will bring me blessings and favor from God. You know my heart and how I have lived my life. I have always tried to do what is right for others. How could I not honor my father and brother as you want? If I have done anything wrong, please tell me, and I will correct it so that God will have mercy on me."

Rebecca said, "My son, I have never seen you do anything wrong, only good. But I must tell you something important. My time is near, and I will die this year. I will not live beyond the age of 155. I have seen it in a dream, and I know it is true."

Jacob laughed at her words because Rebecca was still strong and healthy. She had no sign of illness, moved around easily, and had never been sick. Jacob said, "Mother, if I could live as long as you and still be as strong as you are, I would consider it a blessing. You won't die; you must be joking with me."

Rebecca then went to Isaac and said, "I have a request, my husband. Please make Esau promise that he will not harm Jacob or stay angry with him. You know how Esau has always been—he has been difficult since childhood, and there is no kindness in him. After you pass away, he plans to kill Jacob. You have seen how he has treated us, especially after Jacob left for Haran. He took your flocks and stole from us, and when we asked for what was rightfully ours, he acted like he was doing us a favor. He is still upset because you blessed Jacob, who is honest and righteous. But since Jacob came back, he has cared for us in every way. He shares what he has, respects us, and treats us with kindness."

Isaac replied, "I know everything Jacob has done. He honors us with all his heart. I used to love Esau more because he was my firstborn, but now I love Jacob more because Esau has chosen a sinful path. He does not follow what is right; he has become violent and corrupt. He has turned away from the God of Abraham and follows the ways of his wives, who have led him into impurity. Neither he nor his descendants will be saved; they will be destroyed. As for making Esau promise, even if he does, he will not keep his word. But do not worry about Jacob. He is protected by Someone far greater than Esau's strength."

Rebecca then called Esau to her. When he arrived, she said, "I have a request, my son. Will you promise to do what I ask?"

Esau answered, "I will do whatever you ask, Mother. I will not refuse you."

Rebecca said, "When I die, bury me near Sarah, your father's mother. Also, love your brother Jacob and do not harm him. If you and Jacob love each other, you will both prosper and be honored in this land. No enemy will have power over you, and you will be a blessing to those who love you."

Esau promised, "I will do everything you ask. I will bury you near Sarah when you pass, and I will love Jacob. He is my brother, and it is only natural to love him—we are family. If I do not love him, who else should I love? I also ask that you speak to Jacob about me and my sons. I know he will rule over us, for when my father blessed him, he made him greater and me lesser."

Esau swore to Rebecca that he would do as she asked. Then Rebecca called Jacob to stand before Esau and gave him the same instructions. Jacob replied, "I will do everything you ask. I promise that neither I nor my sons will ever harm Esau. I will only love him."

That evening, they shared a meal and drank together. That night, Rebecca passed away at the age of 155. Esau and Jacob buried her in the cave near Sarah, their grandmother.

Chapter XXXVI.

In the sixth year of that time, Isaac called his two sons, Esau and Jacob, to come to him. He said, "My sons, my time is near. I will soon go to be with our ancestors. When I pass, bury me next to my father Abraham in the cave in the field of Ephron the Hittite—the same tomb Abraham bought for our family. That is where I have prepared my resting place.

I ask you, my sons, to live with honesty and fairness, so that the Lord will fulfill the promises He made to Abraham and his descendants. Love one another as you love your own life. Support each other, work together, and let your bond be strong.

I warn you not to worship idols or be drawn to them. They mislead those who follow them. Remember the Lord, the God of Abraham, and how I served Him with joy and faithfulness. Because of this, He blessed me, made my descendants as numerous as the stars, and established us as a righteous people who will last forever.

Now, I will have you swear a great oath—by the name of the One who created everything in heaven and on earth—that you will honor and worship Him alone. Love your brother with honesty and goodness, and never plan harm against him. If you do this, you will be successful in everything, and no harm will come to you.

But if one of you plots evil against his brother, that person will bring destruction upon himself. He will be cut off from the land of the living, and his family will not survive. When God's wrath comes, as it did in Sodom, his land, city, and everything he owns will be destroyed. His name will be erased from the book of the righteous and placed among those who are condemned. He will face suffering and sorrow forever.

I warn you, my sons: anyone who harms his brother will face judgment. Today, I am dividing my belongings between you. Since Esau is the firstborn, he will receive the larger portion, including the tower and everything Abraham owned at the Well of the Oath."

Esau replied, "I already sold my birthright to Jacob. It belongs to him, and I have no claim to it."

Isaac said, "May God's blessing rest on both of you and your descendants from this day forward. You have brought me peace by making things right between you. My heart is no longer troubled about the birthright. May the Most High bless those who act with righteousness and extend that blessing to their children forever."

After blessing them, Isaac gave his final instructions. They ate and drank together in his presence, and Isaac was filled with joy, knowing his sons were at peace. That night, they went to rest, and Isaac lay in his bed, content. Soon after, he passed away peacefully. He lived to be 180 years old, completing twenty-five weeks and five years. His sons, Esau and Jacob, buried him.

Esau moved to the land of Edom and settled in the mountains of Seir. Jacob remained in the mountains of Hebron, living in the same tower where his grandfather Abraham had once stayed. He continued to worship the Lord with all his heart, following the commands that had been passed down through his family.

In the fourth year of the second week of the forty-fifth jubilee, Jacob's wife Leah passed away. He buried her in the same cave where his mother, Rebecca, had been laid, to the left of Sarah, the mother of his father. All her sons, along with Jacob's other children, came together to mourn her and to comfort Jacob, for he loved her deeply after Rachel had died.

Leah had been kind, faithful, and righteous all her life. She always honored Jacob and never spoke harshly to him. She was gentle, loving, and respectable. Jacob remembered her goodness and mourned her loss with all his heart and soul.

Chapter XXXVII.

On the day Isaac, the father of Jacob and Esau, passed away, Esau's sons found out that Isaac had given the birthright to Jacob, even though Esau was the older son. They were furious and demanded answers from their father.

"Why did your father give the firstborn's blessing to Jacob instead of you?" they asked.

Esau answered, "I sold my birthright to Jacob for a bowl of lentils. Later, when our father sent me to hunt and bring him food so he could bless me, Jacob tricked me. He brought food to our father first and received the blessing that was meant for me. After that, our father made us both swear to live in peace, to love each other, and not to harm one another."

But Esau's sons refused to listen. "We will not make peace with Jacob," they declared. "We are stronger than he is. We will fight him, kill him, and wipe out his sons. And if you refuse to help us, we will deal with you too."

They continued, "Let's send messengers to Aram, Philistia, Moab, and Ammon to gather warriors who love to fight. With their help, we will destroy Jacob before he grows even more powerful."

Esau tried to stop them. "Do not go to war with him," he warned. "If you do, you may be the ones who fall."

But his sons ignored him. "You have spent your whole life obeying Jacob. We will not follow your advice."

Determined to carry out their plan, they sent messengers to Esau's ally, Aduram, and hired a thousand warriors. They also gathered a thousand fighters each from Moab, Ammon, Philistia, Edom, and the Horites. From the Kittim, they brought mighty warriors. Then, turning to Esau, they threatened, "Lead us into battle, or we will kill you."

Filled with anger and frustration, Esau finally agreed. As old feelings of resentment returned, he forgot the oath he had sworn to his parents and allowed his heart to turn against Jacob once again.

Meanwhile, Jacob had no idea that trouble was coming. He was still mourning the death of Leah, his wife, when Esau and his army of four thousand warriors came near the tower where he was staying. The people of Hebron, who respected Jacob more than Esau because of his kindness and generosity, rushed to warn him.

"Esau is coming with four thousand armed men, ready for battle," they told him.

At first, Jacob didn't believe them. But when he saw the army approaching, he quickly shut the gates of the tower and climbed to the top. From there, he called out to Esau.

"Is this how you come to comfort me after my wife's death? Is this how you keep the oath you swore to our father and mother? You have broken your promise and brought judgment upon yourself."

Esau replied bitterly, "Oaths mean nothing. People and animals alike will always fight their enemies. You have hated me and my children for as long as I can remember. We are not brothers, and we never will be."

Then Esau spoke in anger:

"If a wild boar could grow soft fur like a lamb,
If it could sprout horns like a ram or a deer,
Then I would make peace with you.

If a mother could leave her newborn child,
Then I would call you my brother.

If wolves could lie down with lambs,
Without trying to tear them apart,
And their hearts became kind,
Then I would make peace with you.

If a lion and an ox could work together,
Plowing the fields side by side,
Then I would make peace with you.

If a raven could turn as white as snow,
Then you would know that I loved you.

But you and your children will be torn from the land,
And there will never be peace for you.

When Jacob saw the hatred in Esau's heart and the fury in his eyes, he realized Esau was determined to destroy him. Understanding that words would not change his brother's mind, he compared Esau to a wild boar charging straight into a spear.

Jacob then turned to his men and said, "Prepare your weapons. Stand ready. We will not run from this fight."

Chapter XXXVIII.

After that, Judah turned to his father, Jacob, and said, "Father, take your bow and shoot your arrows to defend us. Show your strength, but do not let us harm your brother. He is still your own flesh and blood and should face you in this battle."

Jacob took his bow and fired an arrow, hitting Esau on the right side of his chest, killing him instantly. He shot another arrow, striking 'Adoran the Aramean on his left side, knocking him backward and killing him as well.

Then, Jacob's sons and their servants split into groups and attacked the enemy from all directions around the tower. Judah led the southern group with Naphtali, Gad, and fifty servants. Together, they fought fiercely, leaving no survivors. On the eastern side, Levi, Dan, and Asher, along with fifty men, faced the warriors from Moab and Ammon, defeating them all. Reuben, Issachar, and Zebulon took the north side with fifty men and overpowered the fighters from Philistia. Meanwhile, Simeon, Benjamin, and Enoch, Reuben's son, attacked from the west with fifty men, defeating four hundred warriors from Edom and the Horites.

Even though six hundred men, including four of Esau's sons, managed to escape, they left Esau's body behind on the hill of 'Aduram. Jacob buried Esau there and returned home.

Jacob's sons pursued Esau's fleeing sons into the mountains of Seir, where they captured them and made them serve Jacob's descendants. They sent word to Jacob, asking if they should make peace with Esau's sons or destroy them completely. Jacob instructed them to make peace, so they did, placing Esau's descendants under their rule and requiring them to pay tribute to Jacob and his sons for generations.

Esau's descendants continued to pay tribute until the day Jacob and his family moved to Egypt. Even today, the people of Edom remain under the rule of Jacob's descendants and have never been able to free themselves from this obligation.

These are the kings who ruled over Edom before Israel had any kings:

- The first king was Balaq, son of Beor, and his city was called Danaba.
- After Balaq died, Jobab, son of Zara from Boser, became king.
- When Jobab died, 'Asam from the land of Teman took the throne.
- After 'Asam's death, 'Adath, son of Barad—who had defeated the Midianites in the field of Moab—became king. His city was called Avith.
- When 'Adath died, Salman from 'Amaseqa ruled as king.
- After Salman, Saul from Ra'aboth by the river took the throne.
- Saul was followed by Ba'elunan, son of Achbor.

- When Ba'elunan died, 'Adath became king again, and his wife was Maitabith, daughter of Matarat, granddaughter of Metabedza'ab.

These kings ruled over Edom before any king was established in Israel.

Chapter XXXIX.

Jacob stayed in the land of Canaan, where his father had once lived. These are the events of his family's story.

Joseph, at seventeen years old, was taken to Egypt and sold to Potiphar, a high-ranking officer of Pharaoh and the captain of the guard. Potiphar put Joseph in charge of his entire household, and because of Joseph, the Lord blessed everything in Potiphar's home. Everything Joseph did was successful, and Potiphar noticed that God was with him, making him prosper in all he did. Because of this, Potiphar gave Joseph complete authority over everything he owned.

Joseph was strong and handsome, which caught the attention of Potiphar's wife. She became obsessed with him and constantly asked him to be with her. But Joseph refused. He remembered what his father, Jacob, had taught him about the words of Abraham: that committing adultery with a married woman was a terrible sin, deserving of death according to God's laws. Joseph held on to these teachings and refused to betray his master or sin against God.

For an entire year, Potiphar's wife tried to convince Joseph, but he resisted. One day, when they were alone, she grabbed his coat and tried to force him to be with her. Joseph pulled away, leaving his coat in her hands, and ran out of the house. Furious at being rejected, she decided to take revenge. She screamed and told the household servants that Joseph had attacked her. Later, when her husband came home, she said, "That Hebrew slave you trust tried to take advantage of me. When I screamed, he ran away, leaving his coat behind."

Hearing this, Potiphar became angry. Seeing the coat as proof, he had Joseph thrown into prison, where the king's prisoners were kept. But even in prison, the Lord was with Joseph. The chief jailer noticed how responsible and successful Joseph was in everything he did. Recognizing that God was with him, the jailer put Joseph in charge of all the prisoners, trusting him completely to manage everything well.

Joseph remained in prison for two years. During that time, Pharaoh became angry with two of his officials—the chief cupbearer and the chief baker—and had them imprisoned in the same place as Joseph. The jailer assigned Joseph to take care of them.

One night, both the cupbearer and the baker had dreams that troubled them. They shared their dreams with Joseph, and with God's help, he explained their meanings. Just as Joseph predicted, the cupbearer was restored to his position, while the baker was executed.

Before the cupbearer was released, Joseph asked him to remember him and mention his situation to Pharaoh, hoping to be freed. However, once the cupbearer returned to his position, he completely forgot about Joseph and did not speak of him to Pharaoh. Despite Joseph's kindness, the cupbearer did not remember him at all.

Chapter XL.

One night, Pharaoh had two dreams about a great famine that was coming to the land. When he woke up, he called all the dream interpreters and magicians in Egypt to explain his dreams. But none of them could understand what they meant. Then, the chief butler remembered Joseph and told Pharaoh about him. Joseph was taken out of prison and brought before Pharaoh to hear the dreams.

Joseph told Pharaoh, "Both of your dreams mean the same thing. There will be seven years of great harvests, with plenty of food across Egypt. But after that, seven years of famine will come, so severe that people will forget the years of plenty."

He then advised Pharaoh, "You should put wise and responsible men in charge of storing extra food in every city during the seven good years. That way, when the famine comes, there will be enough food to keep people alive, and Egypt will not be ruined by hunger."

God gave Joseph wisdom and favor in Pharaoh's eyes. Pharaoh said to his servants, "There is no one as wise as Joseph. The spirit of God is with him." So Pharaoh made Joseph the second most powerful man in all of Egypt. He gave Joseph control over the entire land and had him ride in his second chariot. He dressed Joseph in fine clothes, placed a gold chain around his neck, and gave him his official ring as a sign of his authority. A messenger went before him, announcing his new position. Pharaoh told Joseph, "Only I, as king, will be greater than you."

Joseph ruled fairly over Egypt. The officials and workers respected him, and he treated everyone with honesty. He never took bribes or showed favoritism. Because of him, Egypt was peaceful, and God continued to bless him. People admired Joseph, and Pharaoh's kingdom remained well-run and free from trouble.

Pharaoh gave Joseph a new name, Zaphenath-Paneah, and arranged his marriage to Asenath, the daughter of Potipherah, a priest from On. Joseph was thirty years old when he stood before Pharaoh. That same year, Isaac passed away. Just as Joseph had predicted, the land had seven years of plenty. The harvests were so abundant that one portion of seed produced eighteen hundred times more. Joseph collected and stored food in every city until the grain supply was so large that it couldn't even be measured.

Chapter XLI.

During the forty-fifth jubilee, in the second week of the second year, Judah arranged for his oldest son, Er, to marry a woman named Tamar, who came from the daughters of Aram. However, Er did not love her because she wasn't from his mother's Canaanite family. He wanted to marry someone from his mother's people, but Judah refused to allow it. Er was wicked in God's eyes, so God took his life.

After Er's death, Judah told his second son, Onan, to marry Tamar and have children on behalf of his late brother. But Onan knew the children wouldn't be considered his, so he purposely avoided making her pregnant. This angered God, and Onan also died.

Judah then told Tamar to stay in her father's house as a widow until his youngest son, Shelah, was old enough to marry her. However, when Shelah grew up, Judah's wife, Bedsu'el, was against the marriage. In the fifth year of that week, Bedsu'el passed away. The next year, Judah went to Timnah to shear his sheep.

When Tamar heard that Judah was going to Timnah, she took off her widow's clothing, covered her face with a veil, and dressed beautifully. She waited by the road where Judah would pass. When Judah saw her, he thought she was a prostitute and approached her. He said, "Let me be with you." Tamar asked, "What will you give me in return?" Judah replied, "I don't have anything with me right now, but I will leave my signet ring, necklace, and staff as a guarantee until I send payment." Tamar agreed, and they were together. As a result, she became pregnant.

Afterward, Tamar returned home. Later, Judah sent a servant with a young goat to pay her and get his items back, but the servant couldn't find her. The locals told him, "There has been no prostitute here." When the servant returned and told Judah, he said, "Let her keep the items. We don't want to be embarrassed."

Three months later, Judah was told, "Tamar, your daughter-in-law, is pregnant from prostitution." Enraged, Judah went to her father's house and demanded that she be burned as punishment. As she was being taken out, Tamar sent a message to Judah, saying, "The man who owns these items is the father of my child. Do you recognize them?" She showed the signet ring, necklace, and staff. Judah immediately knew they were his and admitted, "She is more righteous than I am. I failed to give her to my son Shelah, as I promised." Judah stopped the punishment, and Tamar's life was spared.

Tamar was never married to Shelah, and Judah never had relations with her again. Later, she gave birth to twin boys, Perez and Zerah, in the seventh year of the second week. Around this time, the seven years of abundance that Joseph had predicted for Egypt came to an end.

Judah deeply regretted what he had done. He realized his mistake and sincerely repented before God. Because of his honest confession, God forgave him, and he never repeated the sin. In a dream, he was assured that his wrongdoing had been forgiven because he had shown true remorse and made amends.

It was also revealed to him that his other sons were not responsible for what had happened with Tamar, and because of this, his family line continued. Judah had followed the strict traditions passed down from Abraham when he first sought to punish Tamar, and his actions later influenced the laws that were established.

Chapter XLII.

In the first year of the third week of the forty-fifth jubilee, a terrible famine spread across the land. The rains stopped, and the ground became dry and lifeless. But in Egypt, there was still food because Joseph had wisely stored grain during the seven good years. As the famine worsened, the Egyptians came to Joseph to buy food, and he opened the storehouses, selling grain in exchange for gold.

Meanwhile, Canaan was struggling. When Jacob heard there was food in Egypt, he sent ten of his sons to buy grain, but he did not let Benjamin go with them. When the brothers arrived in Egypt along with other people looking for food, Joseph recognized them immediately, but they did not know who he was. Acting like a stranger, Joseph accused them of being spies. "Are you here to find weaknesses in our land?" he asked. He then put them in prison for a few days. Later, he released nine of them but kept Simeon as a hostage, telling them to return with their youngest brother to prove they were telling the truth. Without them knowing, Joseph secretly filled their sacks with grain and returned their money.

Back in Canaan, the brothers told Jacob everything that had happened. They explained how the ruler of Egypt accused them of spying and would not release Simeon unless they brought Benjamin back. Jacob was heartbroken. "You have already taken my children from me! Joseph is gone, Simeon is gone, and now you want to take Benjamin too? Everything is against me!" he cried. He refused to let Benjamin go. "His mother had only two sons. One is already gone. If something happens to him, I will be in sorrow for the rest of my life."

When they found their money inside their sacks, they became even more afraid, and Jacob refused to send Benjamin with them. But the famine only grew worse, while Egypt still had plenty of food because of Joseph's careful planning. As their supplies ran low, Jacob told his sons, "Go back and buy more food, or we will starve." But they replied, "We cannot go back unless Benjamin is with us. The man was clear—we must bring him."

Seeing no other choice, Jacob finally agreed. Reuben offered, "Trust him to me. If I don't bring him back, you can take my two sons." But Jacob refused. Then Judah stepped forward and said, "Send him with me. I will take full responsibility. If I don't bring him back, I will carry the blame forever."

At last, Jacob agreed and sent Benjamin with them. He also told them to bring gifts for the Egyptian leader: stacte, almonds, terebinth nuts, and pure honey. On the first day of the second year of the famine, they left for Egypt.

When they arrived, Joseph immediately recognized Benjamin but did not reveal who he was. "Is this your youngest brother?" he asked. "Yes," they answered. Joseph then said, "May the Lord be gracious to you, my son."

Joseph invited them to his house, released Simeon, and prepared a feast. The brothers gave him the gifts, and they all ate and drank together. During the meal, Joseph gave food to each of them, but he gave Benjamin seven times more than the others.

Before they left, Joseph wanted to test them. He told his servant, "Fill their sacks with grain, return their money, and put my silver cup—the one I drink from—in the youngest brother's sack. Then send them on their way."

Chapter XLIII.

The steward followed Joseph's instructions exactly. He filled the brothers' sacks with food, returned their money, and secretly placed Joseph's silver cup in Benjamin's sack. Early the next morning, the brothers left for home. But soon after, Joseph told his steward, "Go after them. When you catch them, ask, 'Why have you repaid kindness with betrayal? You stole my master's special cup!' Bring the youngest back to me right away—I must decide what to do with him."

The steward chased after them, caught up, and repeated Joseph's words. The brothers were shocked and said, "Why would you accuse us of this? We would never steal from your master! We even brought back the money we found in our sacks last time. Why would we take silver from his house? Search our bags! If you find the cup with anyone, let him die, and the rest of us will become your slaves."

The steward replied, "No, only the one who has the cup will stay as my servant. The rest of you may go free."

He searched their sacks, starting with the oldest and ending with the youngest. When he opened Benjamin's sack, there was the silver cup. The brothers were devastated. They tore their clothes in grief, loaded their donkeys, and returned to the city.

When they reached Joseph's house, they fell to the ground before him. Joseph asked, "What have you done? Didn't you think I would find out the truth?"

The brothers answered, "What can we say? How can we prove our innocence? God has exposed our guilt. We are your servants now, including the one who had the cup."

Joseph responded, "I fear God. I will not punish all of you. Only the one who stole the cup will stay as my servant. The rest of you may return to your father."

At this, Judah stepped forward and pleaded, "My lord, please listen. We have an elderly father who loves his youngest son dearly. His life is tied to this boy. If we return without him, our father will die of heartbreak. Please let me stay as your servant instead, and let the boy go home with his brothers. I promised my father I would bring him back safely. If I fail, I will carry this guilt forever."

Joseph could no longer hold back his emotions. Seeing their love for one another, he ordered everyone else to leave the room. Then, with tears streaming down his face, he said in Hebrew, "I am Joseph, your brother."

The brothers were too shocked to speak. Joseph continued, "It's really me—the one you sold into Egypt. But don't be afraid or blame yourselves. God sent me here ahead of you to save lives. This famine has already lasted two years, and there are still five more years without harvests or food. God used me to ensure survival. Hurry back to our father and tell him I'm alive and that God has made me ruler of all Egypt. Bring him and your families here so I can take care of you during the remaining years of famine."

Joseph hugged each of his brothers, crying with them. Then he provided them with wagons, supplies for their journey, fine clothes, and silver. For his father, he sent ten donkeys loaded with the best goods from Egypt, along with grain and bread for the trip.

When the brothers returned to Canaan, they told their father, "Joseph is alive! He is the ruler of all Egypt!" Jacob was stunned and couldn't believe it at first. But when he saw the wagons and all the provisions Joseph had sent, his spirit was lifted. He said, "That is enough! My son Joseph is alive! I will go and see him before I die."

Chapter XLIV.

Israel left Haran on the first day of the third month, beginning his journey to Egypt. On the seventh day, he reached the Well of the Oath and offered a sacrifice to the God of his father, Isaac. As he remembered the dream he had at Bethel, he felt unsure and afraid to continue to Egypt. He thought about sending for Joseph instead, so he wouldn't have to leave Canaan. He stayed there for seven days, seeking guidance and hoping for a sign. During this time, he observed the festival of the first-fruits, even though there was no grain to plant because of the severe famine, which had affected crops, animals, birds, and people alike.

On the sixteenth day of the month, the Lord appeared to Jacob in a vision and called, "Jacob, Jacob." Jacob answered, "Here I am." The Lord said, "I am the God of your fathers, the God of Abraham and Isaac. Do not be afraid to go to Egypt. I will make your family into a great nation there. I will be with you and bring you back safely. You will be buried in your homeland, and Joseph will be with you when you die. Have no fear—go to Egypt."

Encouraged by this, Jacob gathered his sons, grandsons, and belongings. They placed him and all they owned on wagons, and on the sixteenth day of the third month, they left the Well of the Oath. Judah went ahead to meet Joseph and prepare the land of Goshen, which Joseph had chosen as their new home. Goshen was a good place for them because it was fertile and close to Joseph, making it ideal for their livestock and families.

These were the family members who traveled with Jacob to Egypt:

- Reuben, Jacob's firstborn, and his sons: Enoch, Pallu, Hezron, and Carmi—four in total.
- Simeon and his sons: Jemuel, Jamin, Ohad, Jachin, Zohar, and Shaul (whose mother was a Zephathite woman)—seven.
- Levi and his sons: Gershon, Kohath, and Merari—three.
- Judah and his sons: Shela, Perez, and Zerah—three.
- Issachar and his sons: Tola, Phua, Jasub, and Shimron—four.
- Zebulun and his sons: Sered, Elon, and Jahleel—three.

These were the descendants of Leah, along with their sister Dinah, who were born to Jacob in Mesopotamia. Including Jacob himself, thirty members of Leah's family entered Egypt.

From Zilpah, Leah's maidservant:

- Gad and his sons: Ziphion, Haggi, Shuni, Ezbon, Eri, Areli, and Arodi—seven.
- Asher and his sons: Imnah, Ishvah, Ishvi, Beriah, and their sister Serah—six.

Zilpah's descendants who traveled to Egypt numbered fourteen.

From Rachel, Jacob's beloved wife:

- Joseph, who had two sons in Egypt, Manasseh and Ephraim (born to Asenath, daughter of Potiphar, priest of Heliopolis)—two.
- Benjamin and his ten sons: Bela, Becher, Ashbel, Gera, Naaman, Ehi, Rosh, Muppim, Huppim, and Ard—ten.

Rachel's descendants who entered Egypt totaled fourteen.

From Bilhah, Rachel's maidservant:

- Dan and his sons: Hushim, Samon, Asudi, Ijaka, and Salomon—five (though only Hushim survived after arriving in Egypt).
- Naphtali and his sons: Jahziel, Guni, Jezer, Shallum, and Iv—five (Iv was born after the famine years but passed away in Egypt).

Bilhah's descendants totaled twenty-six.

In total, seventy of Jacob's family members traveled to Egypt, including his children and grandchildren. However, five of them—Judah's two sons, Er and Onan, and three others who died without children in Egypt—were buried and counted among the seventy nations of the world.

Chapter XLV.

Israel arrived in Egypt and settled in Goshen on the first day of the fourth month in the second year of the third week of the forty-fifth jubilee. Joseph traveled to Goshen to greet his father, and when they met, he hugged Jacob tightly and wept on his shoulder. Israel said, "Now I can die in peace, because I have seen your face and know that you are alive. Praise the Lord, the God of Israel, the God of Abraham and Isaac, who has shown me mercy and kept His promises. It is enough for me that I have seen you. The vision I had at Bethel has come true. Blessed be the Lord, my God, forever and ever."

Joseph and his brothers sat down and ate together in Jacob's presence. Seeing them reunited, sharing a meal, filled Jacob with great happiness. He thanked God, who had watched over him and kept all twelve of his sons safe.

Joseph arranged for his father, brothers, and their families to live in Goshen, specifically in Rameses and the nearby areas, which were under his control as Pharaoh's ruler. Israel and his family settled in the most fertile part of Egypt. Jacob was 130 years old when he arrived, and Joseph made sure they had enough food throughout the remaining years of famine.

As the famine continued, Joseph collected all the land in Egypt for Pharaoh in exchange for food. He also took the people's livestock and possessions for Pharaoh. When the famine finally ended, Joseph provided the Egyptians with seed in the eighth year so they could plant again. The Nile had finally overflowed its banks, marking the end of the food shortage. During the seven years of famine, the river had failed to flood, only watering the edges of the land. But now, it once again covered the fields, allowing the Egyptians to grow crops and harvest an abundance that year. This was the first year of the fourth week of the forty-fifth jubilee.

Joseph established a law in Egypt that required one-fifth of all harvests to go to Pharaoh, while the remaining four-fifths were for the people to use as food and seed. This law remained in place for generations.

Israel lived in Egypt for seventeen more years, reaching a total age of 147 years, or three jubilees. He passed away in the fourth year of the fifth week of the forty-fifth jubilee. Before he died, he gathered his sons, blessed them, and told them what would happen in Egypt and in the future. He gave each of them a blessing and granted Joseph a double portion of inheritance in the land.

Israel was buried in the double cave in Canaan, near his father Abraham, in the tomb that Abraham had prepared in Hebron. Before his death, Israel gave all his writings and the books of his ancestors to Levi, instructing him to protect them and pass them down through future generations so they would never be lost.

Chapter XLVI.

After Jacob passed away, his descendants thrived in Egypt. Their families grew quickly, and they became a large, united people. The brothers cared for one another, and everyone worked together to support their community. During Joseph's lifetime, they increased greatly in number over ten cycles of seven years. There was no trouble or conflict because the Egyptians respected and valued the Israelites while Joseph was alive.

Joseph lived to be 110 years old. He spent 17 years in Canaan, 10 years as a servant, 3 years in prison, and 80 years as a ruler in Egypt. Before he died, he made the Israelites promise that when they eventually left Egypt, they would take his bones with them. He knew the Egyptians would not allow him to be buried in Canaan.

This was because King Makamaron of Canaan, who was living in

Assyria at the time, had fought against the Egyptian king in a valley and defeated him, forcing him to retreat to the gates of 'Ermon. However, Makamaron was unable to enter Egypt because a new, stronger ruler had taken power. The gates of Egypt were heavily guarded, and no one was allowed to pass through.

Joseph died in the forty-sixth jubilee, during the sixth week, in the second year, and he was buried in Egypt. Eventually, all of his brothers passed away as well, along with their entire generation.

In the forty-seventh jubilee, during the second week of the second year, the king of Egypt went to war against the king of Canaan. Around this time, the Israelites took the remains of Jacob's sons, except for Joseph's, and buried them in the double cave on the mountain. Most of the Israelites returned to Egypt, but a few stayed in the mountains of Hebron, including Amram, your father, who remained with them.

Later, the king of Canaan defeated the Egyptian king and sealed off Egypt's borders. Afterward, he created a harsh plan against the Israelites. He told his people, "The Israelites have grown too large and strong. We must act now before they increase even more. If war breaks out, they might join our enemies and leave Egypt. Their hearts are already set on returning to Canaan."

To control them, he put slave masters over them and forced them to build strong cities for Pharaoh, including Pithom and Raamses. They were also made to repair and strengthen Egypt's cities. The Israelites were treated cruelly, but the more they were oppressed, the more their numbers grew. This made the Egyptians fear and resent them even more, leading to even harsher treatment.

Chapter XLVII.

During the seventh week, in the seventh year of the forty-seventh jubilee, your father left Canaan. You were born in the fourth week, during the sixth year of the forty-eighth jubilee. At that time, the Israelites were suffering greatly. Pharaoh, the ruler of Egypt, had ordered that all newborn Hebrew boys be thrown into the river. For seven months, this cruel law was strictly followed, and many baby boys were cast into the waters.

On the day you were born, your mother hid you for three months to keep you safe. When she could no longer hide you, she made a small basket, sealing it with pitch and tar so it would float. She placed you inside and set it among the reeds along the riverbank. For seven days, she returned at night to nurse you, while your sister Miriam stayed nearby during the day to watch over you and protect you from harm.

One day, Pharaoh's daughter, Tharmuth, came to the river to bathe. She heard your cries and told her maids to bring her the basket. When she saw you inside, she felt compassion and decided to adopt you. Your sister stepped forward and asked, "Shall I find a Hebrew woman to nurse the baby for you?" Tharmuth agreed, and Miriam brought your mother, Jochebed, to care for you. Pharaoh's daughter even paid her to look after you.

When you grew older, your mother brought you back to Tharmuth, who raised you as her own son. Although you were brought up in Pharaoh's palace, your father, Amram, secretly taught you how to read and write, making sure you knew your true heritage. You spent 21 years in the royal court, but one event changed your life forever.

One day, while walking outside the palace, you saw an Egyptian beating one of your fellow Israelites. Overcome with anger, you killed the Egyptian and buried his body in the sand to cover it up. The next day, you saw two Israelites arguing and tried to stop them. You asked the one in the wrong, "Why are you hitting your brother?" But he pushed back and said, "Who made you our ruler or judge? Are you planning to kill me like you killed the Egyptian?"

When you heard this, fear took hold of you. You realized that people knew what you had done, and it would only be a matter of time before Pharaoh found out. Worried for your safety, you fled Egypt to escape the consequences of your actions.

Chapter XLVIII.

In the sixth year of the third week of the forty-ninth jubilee, you left and lived in Midian for five weeks and one year. Then, in the second week of the second year of the fiftieth jubilee, you returned to Egypt. You clearly remember what God told you on Mount Sinai and how Prince Mastêmâ tried to stop you on your way back. He saw that you had been sent to bring judgment on Egypt and used all his power to try and kill you, hoping to prevent you from saving the Israelites. But I rescued you from his grasp, and you carried out the signs and wonders that God had commanded you to perform in Egypt against Pharaoh, his household, his officials, and his people.

The Lord sent powerful judgments against the Egyptians for the sake of Israel. He struck them with plagues: turning the water to blood, covering the land with frogs, tormenting them with lice and gnats, afflicting them with painful boils, killing their livestock, sending hail that destroyed their crops, unleashing locusts that ate whatever was left, covering the land in darkness, and finally, taking the lives of all their firstborn children and animals. The Lord also destroyed their idols, burning them with fire.

Everything happened exactly as you foretold. In front of Pharaoh, his officials, and all of Egypt, you warned them, and each plague came just as you had said. The Lord sent ten devastating plagues to punish Egypt and avenge Israel. He did this to keep His promise to Abraham, repaying the Egyptians for enslaving His people.

But Prince Mastêmâ fought against you the entire time. He tried to hand you over to Pharaoh and helped the Egyptian magicians perform evil tricks. However, we prevented their magic from healing anyone. Instead, the Lord struck them with painful sores so severe that they couldn't even stand, preventing them from performing any more illusions.

Even after witnessing all these miracles, Mastêmâ did not give up. Instead, he encouraged the Egyptians to chase after the Israelites with their full army—chariots, horses, and soldiers. But I stood between them and Israel, protecting My people and delivering them from his hands.

The Lord led Israel safely through the sea, turning the water into dry land so they could cross. But when the Egyptians followed, the Lord threw them into the deep waters, drowning them. Just as they had drowned Israelite children in the river, God repaid them, destroying one million of them, wiping out a thousand strong men for every Hebrew child they had thrown into the water.

On the fourteenth, fifteenth, sixteenth, seventeenth, and eighteenth days, Mastêmâ was bound and held back so he could not accuse the Israelites. On the nineteenth day, we released him, allowing him to influence the Egyptians as they pursued Israel. But God hardened their hearts, making them even more stubborn. This was part of His plan—to bring them to their destruction in the sea.

On the fourteenth day, we bound Mastêmâ so he could not accuse Israel when they took gold, silver, bronze, and clothing from the Egyptians. This was their rightful payment for all the years they had been forced to work as slaves. The Lord made sure the Israelites did not leave Egypt empty-handed.

Chapter XLIX.

Remember the command the Lord gave you about Passover: celebrate it at the right time, on the fourteenth day of the first month. The sacrifice must be made before evening and eaten that same night, as the fifteenth day begins at sunset.

On this special night—when the festival begins and joy fills the hearts of Israel—you were eating the Passover meal in Egypt. That same night, all the forces of Mastêmâ were released to strike down every firstborn in Egypt, from Pharaoh's son to the firstborn of the lowest servant, even the firstborn animals.

The Lord gave His people a sign: any house with the blood of a one-year-old lamb on its doorframe would be protected. The destroyer would not enter but would pass over, sparing everyone inside because of the blood on the door.

The Lord's power worked exactly as He commanded. The plague passed over the houses of the Israelites, leaving them unharmed. No person, animal, or even a dog suffered any harm. But in Egypt, the disaster was severe—every household lost someone, filling the land with mourning and cries of sorrow.

Meanwhile, the Israelites were eating the Passover lamb, drinking wine, and giving thanks to the Lord, praising Him for their deliverance. They were ready to leave Egypt and escape from their suffering.

Remember this day always, and celebrate it every year on the appointed day, following all the instructions. Do not delay or change the date.

This is an everlasting command, recorded in the heavenly books. Every generation of Israel must observe it every year, on the exact date. This law will never change.

Anyone who is able but refuses to celebrate the Passover on the correct day—failing to offer a sacrifice to the Lord and join in the feast—will be cut off from Israel. Because they ignored the Lord's command, they will bear the guilt of disobedience.

The people of Israel must celebrate Passover on the fourteenth day of the first month, from evening to evening, as the day transitions from light to night. The Lord has commanded that it be observed at this specific time, "between the evenings."

The Passover sacrifice must not be made during the day but only at sunset. It must be eaten that night until the first third of the night has passed. Any leftover meat must be burned.

The lamb must not be boiled or eaten raw. It must be roasted over fire with its head, insides, and legs intact. Its bones must not be broken, for just as no Israelite's bones shall be broken, neither shall the bones of the Passover lamb.

The Lord commanded Israel to observe this festival on its exact date, without postponing it. It is a holy day, a time set apart for worship, and must not be rescheduled or moved.

Tell the people of Israel to celebrate Passover every year, as the Lord commanded. It will be a lasting memorial that pleases Him, and no plague will harm those who keep this command.

The Passover meal must not be eaten outside the Lord's sanctuary. All of Israel must come together and celebrate it at the right time.

Every man who is at least twenty years old on the day of Passover must eat it in the Lord's sanctuary, as it is written. They must partake in the feast before the Lord.

When the Israelites enter the land of Canaan—the land given to them as their inheritance—and establish the Lord's tabernacle in the center of the land, in one of their tribes, they must continue celebrating Passover at the tabernacle every year. They must sacrifice the lamb before the Lord as part of their worship.

When the Lord's temple is built in the land, the people must go there to offer the Passover sacrifice at sunset. The lamb's blood must be placed at the altar's entrance, its fat burned on the altar fire, and its meat roasted and eaten in the courtyard of the holy temple.

Passover must not be celebrated in private homes or in different cities. It must only be observed at the tabernacle or the temple where the Lord's name dwells. The people must remain faithful and not turn away from Him.

Moses, instruct the Israelites to follow these Passover commands exactly as I have given them to you. Teach them to celebrate this festival each year and observe the Feast of Unleavened Bread for seven days. During these seven days, they must eat unleavened bread and bring daily offerings before the Lord at His altar.

This festival is a reminder of the night you left Egypt in haste and entered the wilderness of Shur. You completed the celebration by the sea.

Chapter L.

After giving you this law, I also told you about the Sabbath days while you were in the desert of Sin, between Elim and Sinai. I explained the Sabbaths for the land on Mount Sinai, and I also told you about the cycle of jubilee years. However, I did not tell you about the year of the jubilee yet because you will only observe it after entering the land that you will possess. The land itself will also observe the Sabbaths while you live in it, and then you will understand the jubilee year.

For this reason, I have established for you the system of weeks, years, and jubilees. There have been forty-nine jubilees from the time of Adam until today, plus one week and two years. There are still forty more years left for you to learn the commandments of the Lord before you cross over the Jordan River into the land of Canaan. The cycle of jubilees will continue until Israel is completely purified from sin, wrongdoing, and impurity. When that time comes, Israel will live in peace, free from Satan and all evil, and the land will remain pure forever.

I have written down the commandment about the Sabbaths for you, along with all the rules and judgments that come with it. You are to work for six days, but the seventh day is the Sabbath of the Lord your God. On this day, no one should work—not you, your children, your servants, your animals, or any visitor staying with you. Anyone who works on the Sabbath must be put to death.

Anyone who dishonors the Sabbath in any way—by engaging in intimate relations, planning work, starting a journey, buying or selling, drawing water that was not prepared on the sixth day, or carrying a load from their house—must also be put to death. You are to do no work on the Sabbath except what was prepared in advance for eating, drinking, and resting. The Sabbath is a day to bless the Lord, who has given you a special and holy day of rest. It is a day for all of Israel to stop working and observe forever.

The Lord has honored Israel by allowing them to eat, drink, and rest on this festival day, free from the labor of men. The only work that should be done on the Sabbath is offering incense, sacrifices, and offerings to the Lord. These acts of worship must take place in the Lord's sanctuary so that atonement can be made for Israel as a lasting memorial that pleases God. These offerings should be presented to Him daily, as He has commanded.

Anyone who works on the Sabbath, travels, tends to their farm, lights a fire, rides an animal, sails a boat, harms or kills any creature, slaughters an animal or bird, catches any fish, fasts, or goes to war on this day must be put to death. This is so the people of Israel will keep the Sabbath properly, as written in the commandments given to me. These laws were recorded on the tablets, teaching the people about the seasons and how to observe their days.

This completes the instructions about how the days are divided.

Genesis Apocryphon II-XXII

Then I started wondering—was this pregnancy from the Watchers or Holy Ones, or could it be from the Nephilim? I became really worried about this child.

I, Lamech, got scared and went to my wife, Batenosh. I said, "Please tell me the truth by the Most High, by the Lord of Greatness, by the Eternal King—is this child really mine? Is it from the heavenly beings? Don't lie to me. Swear to tell the truth by the Eternal King."

Batenosh, my wife, answered me firmly while crying. She said, "My husband, please remember when I got pregnant. You know we've been together. Can I really explain everything to you?"

That only made me more upset.

When Batenosh saw how my face had changed with worry, she calmed herself down and said, "My husband, remember when I got pregnant. I swear to you by the Great Holy One and by the Ruler of Heaven—this baby is yours. I became pregnant by you. You are the father. This child is not from any outsider, not from a Watcher, not from any heavenly being. Why has your face changed like that? Why do you look so upset? I'm telling you the truth."

Then I, Lamech, rushed to my father, Methuselah, and told him everything. I asked him to go ask his father, Enoch, because Enoch is close to God, and the Holy Ones reveal things to him.

When Methuselah heard everything, he went quickly to Enoch to ask for the truth. Enoch had gone to a place called Parvayyim. When Methuselah found him, he said, "Father, master, I've come to you... Please don't be angry that I came. I have great respect for you."

[The next part is broken, but seems to be about the past during the time of Yared and how things were divided across the earth.]

Later, Enoch said to Methuselah, "My son, this child is from Lamech. He's not from heavenly beings. The baby is different, but he truly comes from your son, Lamech. Lamech was scared of how the baby looked and thought maybe he wasn't the father—but I'm telling you the truth.

Now go back and tell Lamech the truth. Everything that heavenly beings have done on earth has been revealed to me. The baby's face shines like the sun. He is special. Many people will be confused, and God will let them do violent things, but I will tell you the truth to pass on to your son, Lamech. Tell him this mystery. During this child's lifetime, great things will happen. Praise the Lord of all."

When Methuselah heard this, he spoke privately with Lamech and explained everything. Then I, Lamech, understood and believed that the child really was mine.

[Copy of the words of Noah begins here.]

I, Noah, was protected from corruption, even while in my mother's womb. I was meant to be righteous. All my life I followed the path of truth. The Holy One was with me, keeping me away from darkness and lies. I stayed away from violence and lived with wisdom and truth.

When I became an adult, I continued living righteously. I married Amzara, and we had three sons and some daughters. I found wives for my sons from my brother's daughters, and gave my daughters to his sons, just as God had commanded humans to do.

During my life, after ten cycles (jubilees), my sons finished getting married. I had a vision where I was shown what the heavenly beings had done. I kept it secret and didn't tell anyone.

Then one of the great Watchers came to me with a message from the Holy One. He spoke to me in a vision and said, "Noah, hear this message..." I already knew what people on earth were doing, and I explained it clearly. The Watcher told me that blood had been spilled by the Nephilim. I stayed silent and waited.

The Holy Ones who had children with human women caused many problems. But I, Noah, found favor and righteousness in God's eyes. I saw the gates of heaven and learned many things—not just about people, but animals, birds, and more.

I looked at the land and everything on it—the seas, the mountains, the stars, sun, moon, and the Watchers. God handed all of it over to me. I rejoiced at the words of the Lord of Heaven and called others to hear.

[Several lines are missing here.]

I gathered everything. My wife came with me. We prepared for what would come next.

[Large sections are missing.]

Then came the time when my sons and I gave thanks and blessings to God at night. We praised the King of All Ages, forever and ever.

[Missing section.]

Eventually, the ark settled on one of the mountains of Ararat. Then I made an offering. First, I sacrificed a goat and burned its fat. Then I poured its blood at the base of the altar and burned its meat. I also offered doves and placed flour mixed with oil and incense as a meal offering. I added salt to all of it, and the sweet smell of the offering rose to heaven.

Then the Most High responded...

[The rest of the text is mostly missing.]

I, Noah, was at the doorway of the ark...

[Several lines are missing again.]

Then I, Noah, traveled across the land, walking through its length and width. Everywhere I looked, trees were full of leaves and fruit. The ground was covered in plants, grass, and grain. I praised the Lord of Heaven, who had done good things. He is eternal and worthy of praise. I thanked Him again for having mercy on the earth—He removed all the violent and wicked people but saved a good man because of his righteousness.

Then the Lord of Heaven appeared to me. He spoke and said, "Noah, don't be afraid. I will stay with you and with your children, those who are like you, forever. I will give you and your descendants rule over the land, its deserts, its mountains, and everything in them. I give all of it to you so you can eat its vegetables and plants—but you must not eat blood. Everyone will fear you. I belong to you."

[Part missing.]

God said, "Look, I have placed my rainbow in the clouds. It will be a sign for me and for the earth."

Later, while in the mountains, I planted a vineyard on the slopes of Mount Ararat. After that, I, along with my sons and grandchildren, went down to lower ground. The world had been deeply damaged. Sons and daughters were born to me after the flood. My oldest son Shem had a son named Arpachshad two years after the flood. Shem's other sons were Elam, Asshur, Lud, and Aram. He also had five daughters.

Ham's sons were Kush, Mizraim, Put, and Canaan. He had seven daughters.

Japheth's sons were Gomer, Magog, Madai, Javan, Tubal,

Meshech, and Tiras. He had four daughters.

Together with my sons, I started farming the land. I planted a big vineyard on Mount Lubar. After four years, it gave me wine.

During the first harvest festival in the seventh month, I opened a container of wine and drank from it. It was the first day of the fifth year. That day, I called my sons, grandsons, our wives, and their daughters. We gathered together and celebrated. I praised the Lord of Heaven, the Most High God, the Great Holy One who saved us from destruction.

[Lines missing.]

Later, I noticed something strange—wild animals and insects changed, and people began taking gold, silver, and other things for themselves. I watched them cut down trees and take them. Even the sun, moon, and stars seemed to be involved. I watched until sea and land animals destroyed things, and it all came to an end.

I turned to see the olive tree. It grew tall and had many leaves. I studied it carefully. It had so many leaves and branches tied together. Then strong winds from the four directions blew on it and tore off its branches. The wind from the west hit first, shaking it and scattering leaves and fruit everywhere.

[More lines missing.]

Then someone said, "Pay attention and listen! You are like a large cedar tree. In a dream, you saw a shoot growing from it—that means your three sons. One of those sons is closely connected to you and will stay connected to you for his whole life. Your name will continue through his children, and from him will come a righteous family that will last forever."

"You saw another shoot from the cedar... and another... and some of their branches tangled together, showing how two sons are connected."

[More lines missing.]

Then someone came from the south holding a sharp tool with fire. He came to destroy wickedness. He would burn everything evil. Four angels also appeared. They went among all the nations. Everyone became confused and started serving them. But don't be surprised—everything written about you is true.

Then I, Noah, woke up. The sun was shining. I remembered what I saw about the righteous person.

[More missing lines.]

My son Shem divided his land among his sons. Elam got the land north of the Tigris River, which reaches the Red Sea. Asshur's land went west to the Tigris. Aram got the land between two rivers—Mesopotamia—up to Mount Asshur. Arpachshad received land watered by the Euphrates and the valleys and islands nearby. This was the land I, Noah, gave him.

Japheth also divided land among his sons. Gomer got the north up to the Tina River. Magog, Madai, Javan, Tubal, Meshech, and Tiras each received land and coastal areas near Lydia and other inlets. Their territories stretched toward the land of Ham's children.

[Column mostly missing.]

Later, I built an altar and called on the name of God. I said, "You are my God forever." I hadn't reached the holy mountain yet. I was still traveling south. Eventually, I came to Hebron, which had already been built. I lived there for two years.

There was a famine in the land, but I heard there was food in Egypt. So I moved toward Egypt. I reached the Carmona River, one of the Nile's branches. We left our land and entered the land of Ham's descendants—Egypt.

That night, I had a dream. In it, I saw a cedar tree and a palm tree. The palm tree was beautiful. People came to cut down the cedar but the palm stopped them. She said, "Don't cut down the cedar—we come from the same root!" Because of her, the cedar was saved.

That night, I woke up from a dream and told my wife Sarai, "I had a dream, and it really scared me." She said, "Tell me what you saw so I can understand it." I told her that people would try to kill me but would let her live. I said, "Please, wherever we go, say that I'm your brother. That way, I'll be safe. They might try to take you and kill me!" When she heard this, Sarai started crying.

Later on, Pharaoh Zoan wanted Sarai to come with me to his city. Sarai was very careful to stay out of sight so no one would notice her. But after five years, three Egyptian princes came to Pharaoh and asked about Sarai and me. They gave me many gifts and asked if I could teach them about wisdom, truth, and good values. I read to them from the Book of Enoch. Even though there was a famine, people were still eating and drinking a lot.

The princes talked about how beautiful Sarai was. They admired everything about her—her face, her eyes, her hands, her voice. They said no one else could compare to her. Not only was she beautiful, but she was also wise and talented. When Pharaoh heard their words, he fell in love with Sarai and quickly sent people to bring her to him. When he saw her for himself, he was amazed by her beauty and decided to marry her. He wanted to kill me, but Sarai said, "He is my brother," so that I would be safe. Because of her, my life was spared.

That night, I cried deeply. My nephew Lot cried with me when Sarai was taken away. I prayed to God with all my heart. I said, "You are the Lord of all time and all kings. Please help me! Pharaoh took my wife by force. Show Your power and stop him. Let everyone know You are in charge." Then I stopped praying.

That same night, God sent a harmful spirit to Pharaoh's house. Everyone there became very sick, including Pharaoh. He couldn't go near Sarai, and he didn't touch her. This continued for two years. As things got worse, Pharaoh brought in his magicians, doctors, and wise men to try to help. But none of them could heal him because the spirit attacked them too, and they ran away.

Then a man named Harqanosh came to me. He had seen me in a dream and asked me to pray for Pharaoh. But Lot said, "My uncle Abram can't pray while Sarai is still with the king. Tell Pharaoh to give her back. Then Abram will pray, and the king will be healed."

So Harqanosh told Pharaoh, "All this sickness is because of Sarai, Abram's wife. Give her back, and everything will be fine." Pharaoh called me and said, "Why did you lie to me? You said she was your sister, but she's your wife! Take her and leave Egypt! But please pray for me and my family so the spirit will leave us."

I prayed, placed my hands on Pharaoh's head, and the spirit left him. He got better. He gave me gifts and promised he had not touched Sarai. He gave her back to me with silver, gold, fine clothes, and a servant named Hagar. He also sent people to guide us out of Egypt.

I left Egypt with many riches. Lot was with me, and he had also gained a lot. He married an Egyptian woman. We traveled back through the places we had stayed until we reached Bethel. There, I rebuilt the altar and offered sacrifices to God. I thanked Him for keeping us safe and giving us so much.

Later, Lot and I had to go separate ways because our shepherds were arguing. Lot went to live in the Jordan valley and settled in Sodom. I stayed in the hills near Bethel. I felt sad that Lot had left.

That night, God appeared to me in a dream and said, "Go to Ramath-Hazor, north of Bethel. Look in every direction. I will give all this land to you and your family forever. I will make your family as many as the dust on the ground—too many to count. Walk through the land and see it, because it will belong to you." The next day, I went to Ramath-Hazor and looked around. I saw all the land from Egypt to Lebanon, from the Mediterranean to the desert near the Euphrates. God promised it all to me.

I traveled across the land—from the Gihon River to Mount Taurus, then along the east side of the Salt Sea to the Euphrates and the Red Sea. I followed the coastline until I reached the Sea of Reeds. Then I turned south and returned home safely, where I found everything and everyone in good shape.

I settled near the oaks of Mamre by Hebron and built another altar. I offered sacrifices again. My whole household celebrated, and I invited my friends—Mamre, Arnem, and Eshkol, three Amorite brothers—to eat and drink with us.

Before all this, Kedorlaomer, the king of Elam, had teamed up with other kings to fight against Sodom and nearby cities. They won and made the cities pay them for twelve years. In the thirteenth year, the cities rebelled. In the fourteenth, Kedorlaomer came back and attacked many places. He defeated the Rephaim, Zumzamim, Emim, and others, and eventually captured Sodom and took Lot and all his things.

One of Lot's shepherds escaped and came to tell me. I cried for Lot but then gathered 318 trained men and went after the kings with my friends Arnem, Eshkol, and Mamre. We found them near Dan and attacked them at night from all sides. We beat them and chased them as far as Helbon near Damascus. I rescued Lot, all the captives, and everything they had stolen.

The king of Sodom came to meet me in Jerusalem, also called Salem. I was camped in the Valley of Shaveh. Melchizedek, the king of Salem and a priest of God Most High, brought us food and drink. He blessed me, saying, "May God bless you, Abram. Praise God, who gave you victory!" I gave him a tenth of what I had recovered.

The king of Sodom said, "Give me the people you rescued. You can keep everything else." I told him, "I swore to God that I wouldn't take even a shoelace from you. I don't want you saying you made me rich. The only exceptions are what my men ate and the shares for my three companions—let them decide what to do with theirs."

I returned all the possessions and freed all the captives. Later, God appeared to me in another vision and said, "It's been ten years since you left Haran—two here, seven in Egypt, one since your return. Look at all you have now. It's doubled! Don't be afraid. I'm with you. I'll protect you, support you, and make your wealth even greater."

But I said, "Lord, I have all this wealth, but what's the point if I have no children? When I die, my servant Eliezer will inherit everything." And God answered, "No, Eliezer won't be your heir. You will have a son, and your own child will inherit everything."

Testament of Amram

Introduction

Among the ancient writings found in Cave 4 at Qumran, researchers uncovered a damaged Aramaic document that gives a different view of Amram, the father of Moses. This text, known as the Visions of Amram, tells a story that does not completely match the one in the Bible. It blends historical events with supernatural experiences. While it has some similarities to the Book of Exodus, it seems to be more of a unique interpretation rather than a direct account of biblical history.

One of the most interesting parts of the text is the mention of Amram's age when he died—137 years—just like it is written in Exodus 6:20. The document also gives a different timeline for how long the Israelites lived in Egypt, saying they were there for 152 years. This is much shorter than the 400 or 430 years mentioned in Genesis 15:13 and Exodus 12:40. Some experts, including J. Heinemann, have studied this shorter timeline and connected it to the idea that the Israelites were in Egypt for 210 years (JJS 22, 1971).

Experts who have studied the handwriting believe that different copies of the text were written between the second and first centuries BCE. The fragments, labeled 4Q543–549, show variations in script style, which helps estimate their age.

One of the most mysterious parts of the text describes a vision Amram had. In it, he meets two powerful supernatural beings: the Angel of Darkness, called Melkiresha, and the leader of the Army of Light. The name of this second figure is missing due to damage in the manuscript, but many scholars believe it could be Melchizedek, a mysterious priest-king mentioned in both biblical and other ancient writings.

The Testament of Amram

This is a copy of the book containing the words from the vision of Amram, son of Kehat and grandson of Levi. It includes everything he told his sons and the instructions he gave them on the day he died, at the age of 137. This was also the 152nd year since the Israelites had been exiled in Egypt.

Amram called for Uzziel, his youngest brother, and arranged for him to marry his daughter, Miriam. He told Miriam, "You are thirty years old." To celebrate, he held a feast that lasted seven days. During this time, he ate, drank, and rejoiced. When the feast was over, he sent for his son Aaron, who was about twenty years old, and told him, "Go, my son, and call the messengers, your brothers, from the house of ..."

Qahat went to live and settle in a new place, working alongside many of his relatives. Their work was difficult and continued until they had buried the dead.

In the first year of my journey, I heard troubling news about an upcoming war. With my approval, our group returned to Egypt, and I went to bury the dead. However, they did not build the tombs of our ancestors. My father, Qahat, and my wife, Jochebed, left me behind to continue working, providing them with everything they needed from the land of Canaan. We stayed in Hebron while we built.

A war broke out between the Philistines and the Egyptians. The Philistines and Canaanites defeated Egypt and blocked its borders. Because of this, my wife Jochebed was unable to leave Egypt and travel to Canaan for forty-one years. We could not return to Egypt either. The war between Egypt, Canaan, and the Philistines prevented us from being together.

During all this time, my wife remained in Egypt, separated from me. I did not take another wife. Other women were available, but I refused, hoping to return to Egypt and see my wife again in peace.

One night, I had a vision in a dream. I saw two powerful beings arguing about me. They were having a heated debate over me, so I asked them, "Why are you arguing about me?"

They replied, "We have been given control over all humans." Then they asked me, "Which one of us do you choose?"

I looked up and saw one of them. His appearance was terrifying, like a viper. His clothes were made of many colors, and his skin was extremely dark.

Then I looked at the other one. His face also resembled a snake, possibly an adder. He was covered with something strange, and over his eyes, there was…

Fr. 2

I asked about the Watcher, "Who is he?"

The other one answered, "This Watcher has three names: Belial, Prince of Darkness, and Melkiresha."

Then I asked, "My lord, what power does he have?"

He replied, "All his ways are filled with darkness, and everything he does is evil. He exists in the shadows, and he controls all that is dark. But I rule over all that is light and everything that belongs to it…"

This is a copy of the written words from the vision of Amram, son of Qahat and grandson of Levi. It contains everything he told his sons on the day he died, in the year of his death at the age of 136.

This was also the 152nd year since the Israelites had been exiled in Egypt.

During this time, Amram called for his younger brother, Uzziel, and gave him Miriam, his daughter, in marriage. Miriam was thirty years old. To celebrate, he held a wedding feast that lasted seven days. He ate, drank, and rejoiced throughout the celebration. When the feast ended, he sent for his son Aaron, who was about twenty years old, and told him, "My son, call the messengers, your brothers, from the house of ..."

Amram then spoke and said, "I will explain to you the meaning of your names, just as he wrote for Moses. I will also reveal the mystery of Aaron's service to God. He is a holy priest of the Most High, and all his descendants will remain holy for all generations to come. He will be known as the seventh among the men chosen by God. He will be called by this title and will be chosen as a priest forever..."

I am telling you the true path. I will share this knowledge with you.

All the Sons of Light will shine, while the Sons of Darkness will remain in darkness. The Sons of Light will grow in wisdom and understanding, but the Sons of Darkness will lose their way. In the end, the Sons of Darkness will be removed.

Fools and wicked people will live in darkness, but those who are wise and righteous will shine. The Sons of Light will move toward the light, while the Sons of Darkness will face death and destruction.

The people will be surrounded by brightness, and they will be taught the truth.

Enoch II
The Book of the Secrets of Enoch
(Slavonic Enoch)

Short Account of The Book

The book known as The Secrets of Enoch has only been preserved in the Slavonic language. To keep things simple, we will call it Slavonic Enoch to distinguish it from an older book of Enoch, which has been fully passed down in the Ethiopic language. For convenience, we will refer to that older text as Ethiopic Enoch.

This newly discovered Enoch text was found in recent times through manuscripts in Russia and Serbia. I first became aware of it while working on the Ethiopic Enoch. In an article by Kozak about Russian Pseudepigraphic Literature (published in 1892), I read that a Slavonic version of the Book of Enoch existed, even though only the Ethiopic version had been known before. I quickly reached out to Mr. Morfill for help, and within a few weeks, we obtained printed copies of two of the manuscripts mentioned. After careful study, we realized that Kozak's claim was incorrect.

Instead of being just another version of the Ethiopic Enoch, The Secrets of Enoch turned out to be an entirely different book. However, it is just as important in many ways. Slavonic Enoch was written around the beginning of the Christian era, and its author (or final editor) was a Hellenistic Jew. The book was composed in Egypt.

Because of when and where it was written, it likely did not have a direct impact on the writers of the New Testament. However, its language and ideas are surprisingly similar to some parts of the New

164

Testament, even helping to explain certain difficult passages. Although knowledge of this book was lost for nearly 1,200 years, it was widely read by early Christians and even some groups considered heretical.

Some sections of this book, though often not credited, appear in other ancient writings, such as the Book of Adam and Eve, the Apocalypses of Moses and Paul (written around 400–500 AD), the Sibylline Oracles, the Ascension of Isaiah, and the Epistle of Barnabas (written around 70–90 AD). It is even mentioned by name in the apocalyptic sections of the Testaments of Levi, Daniel, and Naphtali (around 1 AD). Origen referred to it, and it was likely known by Clement of Alexandria and used by Irenaeus. Some phrases from the New Testament may have even been influenced by it.

The Slavonic Manuscripts

The Slavonic version of the Book of Enoch has been translated into English for the first time here. It exists in two main versions, both of which come from a lost Greek original. The surviving manuscripts can be grouped into different categories.

The first category includes complete versions of the text. Two such manuscripts exist. One, owned by Mr. A. Khludov, is a South Russian version from the late 1600s. It is part of a collection containing religious writings, including the lives of saints. Mr. A. Popov published this text in 1880 in the Transactions of the Historical and Archaeological Society of the University of Moscow. However, this manuscript contains many errors. While it serves as the foundation for this translation, I have made corrections using other sources. In this translation, it is labeled as "A."

Another complete manuscript was discovered in 1886 by Professor Sokolov of Moscow in the Public Library of Belgrade. This version is Bulgarian, with a writing style typical of Middle Bulgarian, and it likely dates to the 1500s. It contains extra material, including stories about Methuselah's priesthood, Nir, Melchizedek's birth, and the Great Flood. These additions are not originally part of the Book of Enoch but appear as an appendix.

There is also a shorter and incomplete version of the text, known from three manuscripts. One is housed in the Public Library of Belgrade, a Serbian version published by Novakovic in 1884 in the 16th volume of the literary magazine Starine. This manuscript, dating to the 1500s, has some unique readings and is referred to as "B" in this translation. Another similar manuscript is kept in the Vienna Public Library, and a third, from the 1600s, belongs to Mr. E. Barsov of Moscow.

Of these manuscripts, I have direct access only to "A" and "B." My understanding of the others comes from Professor Sokolov's edited text, which includes all the manuscripts. However, his edition does not clearly separate them. To avoid confusion, I refer to his combined text as "Sok," meaning it represents all sources other than "A" and "B."

Additional fragments of the Book of Enoch can be found in Tikhonravov's Memorials of Russian Apocryphal Literature and Pypin's Memorials of Old Russian Literature. References in early Slavonic writings suggest that these later manuscripts are copies of much older ones that no longer exist. For example, Tikhonravov mentions a manuscript from the 1300s.

My main goal as a translator has been to create a version useful to Western scholars studying apocryphal literature. I have not focused on linguistic details, so my Slavonic colleagues should not criticize me for avoiding in-depth language analysis. That type of study is not yet a priority in England. Instead, my translation aims to support my friend, the Rev. R. H. Charles, so he can explore this subject more thoroughly in relation to Biblical apocryphal literature.

I would also like to thank Professors Sokolov and Pavlov of the University of Moscow. Professor Sokolov kindly let me use his revised text and provided helpful notes on difficult sections. Professor Pavlov has shown great interest in this work, and I am grateful for his support. With the help of Mr. Charles and others, I am happy to have made a small contribution to this field of study.

The Son of Ared;
A Man Wise and Beloved of God

[Concerning the Life and the Dream of Enoch]

There was once a very wise man who achieved great things. God loved him deeply and chose him to see the heavenly realms, the places of wisdom, and the eternal, unchanging God. He was shown the Lord of all—glorious, bright, and beyond imagination. He saw the shining presence of God's servants, the throne that no one can approach, and the countless spiritual beings who serve the Lord. He witnessed their different forms, heard their indescribable songs, and saw the vastness of the universe.

At that time, he said, "When I was 165 years old, my son Methuselah was born. After that, I lived another 200 years, making my total lifespan 365 years."

On the first day of the first month, I was alone in my house. I lay down on my bed and fell asleep. As I slept, a deep sadness filled my heart, and I began to weep in my dream. I didn't understand why I was feeling this way or what was about to happen to me.

Then, two men appeared before me. They were very tall, unlike any humans I had ever seen. Their faces shone like the sun, their eyes burned like torches, and fire came from their mouths. Their clothes looked like they were made of feathers, their feet glowed purple, and their wings were brighter than gold. Their hands were as white as snow. They stood by my bed and called me by my name.

I woke up and saw them clearly standing in front of me. I was terrified and quickly bowed before them. My face changed because of the fear I felt.

The men said to me, "Do not be afraid, Enoch. Be strong. The eternal God has sent us to you. Today, you will go with us into heaven."

They continued, "Tell your sons, your servants, and everyone in your household not to search for you until the Lord returns you to them."

I quickly did as they said and left my house. I called my sons—Methuselah, Regim, and Gaidal—and told them everything these two men had said to me.

Chapters I. 2 II.2

Listen to me, my children, because I do not know where I am going or what will happen to me. I ask you now, my children, do not turn away from God. Stay faithful to the Lord and follow His ways. Do not worship useless idols that did not create the heavens and the earth, because they will be destroyed along with those who follow them.

May God give you strength to remain faithful and always respect Him. And now, my children, do not search for me until the Lord returns me to you.

After I finished speaking to my sons, the two men who had appeared to me called me to them. They lifted me up on their wings and placed me on the clouds. The clouds began to rise, carrying me higher and higher.

As I ascended, I looked down and saw the air far below me. As we went even higher, I entered a new place beyond the sky. The men brought me to the first heaven, where they showed me a vast and powerful sea—much larger than any sea on earth.

Chapters II.3VII. 1

They brought me before the elders and leaders in charge of the stars, showing me two hundred angels who guide the stars and make sure they follow their paths in the sky. These angels had wings and moved in circles around the stars, keeping them in their correct places.

Then I looked and saw huge storehouses filled with snow and ice, guarded by angels who watched over these powerful treasures. I also saw where the clouds were kept, the place where they form and where they return.

Next, they showed me the storehouses of the dew, which looked like a fine, anointing oil. Its colors were as beautiful and varied as all the colors on earth. Many angels were assigned to guard these places, opening and closing them at the right times.

The men who were guiding me then took me to the second heaven. There, I saw a dark and dreadful place where prisoners were hanging, waiting for eternal judgment. These angels were filled with sorrow and despair, darker than the deepest shadow on earth. They cried out constantly, their voices filled with pain.

I asked the men with me, "Why are these beings suffering endlessly?" They answered, "These are the ones who turned away from the Lord. They refused to follow His commands and chose their own ways instead. They rebelled with their leader, and now they are imprisoned here in the second heaven."

I felt deep pity for them. Then, to my surprise, the angels turned toward me, bowed, and said, "Man of God, pray to the Lord for us." But I answered, "Who am I, a mere human, to pray for angels? I don't even know where I am going or what will happen to me. I don't even know who could pray for me."

Chapers VI I. 2V III. 5

[Of the taking of Enoch to the third Heaven 2.]

The men then took me to the third heaven. They placed me in the middle of a breathtaking garden, a place more beautiful than anything I had ever seen.

I saw trees of every color, full of ripe, sweet-smelling fruit that filled the air with a wonderful fragrance. There was an endless supply of food, each kind giving off its own pleasant scent.

At the center of the garden stood the Tree of Life, where God Himself rests when He visits Paradise. This tree was beyond words, its beauty unmatched. It gave off a heavenly fragrance, and its appearance shone like gold, deep red, and something like glowing fire, radiating light all around.

From its roots, four streams flowed—one of honey, one of milk, one of oil, and one of wine. These streams moved gently, spreading out in four directions. They flowed toward the Paradise of Eden, existing between the worlds of the living and the eternal, and continued their course across the earth, moving in harmony with the rest of creation.

There was also an olive tree that never stopped producing oil. Every tree in this garden bore fruit, and each one was overflowing with blessings.

Three hundred glorious angels guarded the garden, singing songs of praise without end. Their voices filled the air, offering worship to the Lord every day. I was amazed by the beauty of this place and said, "What an incredibly blessed place this is!" The men who were with me answered.

Chapters VIII. 6X. 2

[The showing to Enoch of the Righteous, and the Place of Prayer]

"This place, Enoch, has been prepared for the righteous. It is for those who have endured many hardships and attacks in their lives, yet have remained strong. It is for those who turn away from evil and choose to do what is right. It belongs to those who feed the hungry, clothe those in need, help the weak, and care for orphans who have no one to protect them. It is for those who live blamelessly before the Lord and serve Him with all their hearts. This is their eternal reward, a place of peace prepared just for them."

Then the men took me to the northern side, where I saw a place so terrifying that words could not describe it. It was filled with unbearable suffering. Thick darkness covered everything, and it was impossible to see through the heavy gloom. There was no light, only a fierce and never-ending fire. A river of flames flowed through it without stopping. The entire place was surrounded by burning fire, yet at the same time, there were icy winds and freezing cold. It was a place of both unbearable heat and bitter cold.

Inside this terrible place were prisoners who looked wild and tormented. The angels guarding them were fierce and showed no mercy. They carried terrifying weapons and punished the prisoners without relief. I cried out, "Oh no! This place is horrifying!"

The men with me answered, "This place, Enoch, is for those who have rejected God and lived wickedly on earth.

"It is for those who have committed terrible sins, practiced witchcraft, and used dark magic. It is for those who took pride in their evil actions, stole from others, spread lies, and caused harm out of jealousy. It is for those who lived in impurity, committed murder, and took advantage of the weak.

"This place belongs to those who let the hungry starve when they could have fed them and left the poor without clothing when they had the power to help. It is for those who did not acknowledge their Creator but instead worshiped idols—lifeless objects that cannot see or hear. These false gods were made by human hands, yet they bowed down to them as if they had power.

"For all these people, this place has been prepared as their eternal punishment."

Chapters X.3XI.1

[Here they took Enoch to the fourth Heaven, where is the Course of the Sun and Moon.]

The men then took me to the fourth heaven, where they showed me how the sun and moon move along their paths. I saw how bright their light was and measured their courses. I learned that the sun's light is seven times stronger than the moon's. I watched as they traveled in their orbits, moving quickly like a rushing wind. They never stopped, continuing their journey day and night without rest.

I saw four large stars moving alongside the sun. Each of these stars had a thousand smaller stars following on its right side, and another four stars had a thousand stars each on their left side. In total, there were eight thousand stars surrounding the sun.

Around the sun, I saw an enormous group of angels—fifteen myriads of them—who traveled with it during the day, guiding and watching over it. At night, a thousand angels took their place to accompany it.

Each of these angels had six wings, and they flew in front of the sun's chariot, leading and directing its path. A hundred other angels were given the task of keeping the sun warm and bright, making sure its light and heat reached the earth. It was an incredible sight, a breathtaking display of heavenly order and divine power.

Chapters XI. 2—XII.I

[Of the wonderful Creatures of the Sun.]

I looked and saw amazing flying creatures, unlike anything I had ever seen before. They were called phoenixes and chalkadri, and their appearance was both incredible and strange. These creatures had the feet and tails of lions, but their heads looked like crocodiles. Their bodies shimmered with a purple glow, shining like a bright rainbow. Each one was enormous, measuring nine hundred units in size.

They had wings like angels, with twelve large and powerful wings each. These creatures served the sun's chariot, traveling alongside it on its journey. They carried out tasks given to them by God, bringing heat and dew to the earth as commanded.

As the sun moved along its path, these creatures followed, guiding its way beneath the sky and through the hidden places under the earth. The sun's light never stopped shining, constantly moving to brighten the world and sustain life below.

Chapters XII. 2XIII. 5

[The Angels took Enoch, and placed him on the East at the Gates of the Sun.]

The men took me to the East, where they showed me the gates

through which the sun rises at different times. The sun follows a set pattern based on the changing seasons, the months of the year, and the hours of day and night.

I saw six enormous gates, each carefully measured. They were massive, with each gate measuring sixty-one stadia and a quarter of a stadium. I measured them myself and confirmed their enormous size. These gates are where the sun begins its journey, moving westward and following a path that changes with the months and seasons.

The sun travels through the first gate for forty-two days, then through the second gate for thirty-five days. It moves through the fourth and fifth gates for thirty-five days each. However, when it passes through the sixth gate, it stays there for forty-five days.

After that, the sun reverses its path, returning from the sixth gate. It moves through the fifth gate for thirty-five days, then the fourth gate for thirty-five days, followed by the third gate for thirty-five days, and finally through the second gate for another thirty-five days.

This completes the full year, with all the days perfectly matching the sun's cycle and the changes in the four seasons.

Chapters XIV. 1XV. 4

[They took Enock to the West.]

The men then took me to the western part of the sky and showed me six huge open gates. These gates were just like the ones in the East, and through them, the sun sets after completing its journey across the 365 days and a quarter of a day in a year.

As the sun passes through the Western gates, 400 angels come to take its crown and bring it before the Lord. The sun, traveling in its chariot, remains without its light for seven full hours during the night. But when it reaches the East during the eighth hour, the 400 angels return its crown and place it back on the sun.

At that moment, the Phoenixes and Chalkidri, special creatures of the sun, begin to sing. Because of their song, all the birds in the world flap their wings with joy, praising the giver of light. These creatures sing at the Lord's command.

The sun then rises again, spreading its light across the entire earth. The men explained to me how the sun's movements are measured and showed me the gates through which it enters and exits. These great gates were designed by God to mark the passing of days and to set the cycle of the year.

This is why the sun is so large and plays such an important role in the order of the world.

Chapter XVI.17

[The Men· took Enoch and placed him at the East, at the Course of the Moon.]

The men then explained to me how the moon's movements are calculated. They showed me its paths and cycles, pointing out twelve large gates stretching from west to east. The moon enters and exits through these gates at its set times.

The moon moves through the first gate when the sun is in the west, staying there for exactly thirty-one days. It also spends thirty-one days in the second gate. In the third and fourth gates, it remains for thirty days each. It continues this pattern, staying in the fifth and sixth gates for thirty-one days, in the seventh gate for thirty days, in the eighth and ninth gates for thirty-one days, in the tenth gate for thirty days, in the eleventh for thirty-one days, and in the twelfth for twenty-eight days. This cycle repeats as the moon moves through the western gates, following the same pattern as the eastern gates, completing the year.

The sun's yearly cycle is 365 days and a quarter, but the lunar year is only 354 days, made up of twelve months of twenty-nine days each. This leaves an extra eleven days that must be added to match the sun's full cycle for the year. These additional days, called lunar epacts, make up the difference between the sun and the moon's cycles.

Over three years, the extra quarter day from each year is ignored, but in the fourth year, the missing time is accounted for. That is why three years seem to be missing days, but in the fourth year, everything aligns again. To keep the calculations correct, two extra months are added over time, while others are slightly adjusted to maintain balance.

Once the moon completes its cycle through the western gates, it returns to the eastern gates, bringing its light. It moves constantly, both day and night, traveling along its set path faster than the winds of the sky. Alongside it are spirits, creatures, and angels, each with six wings, who guide its movements.

Seven months of the moon's cycle are also measured within a larger cycle of nineteen years, ensuring that everything in the sky stays in perfect order.

Chapters XVI. 8XVIII. 3

In the middle of the heavens, I saw a vast army of angels, armed and ready to serve the Lord. They played cymbals and organs, and their voices rose in endless songs of praise. The sound was unlike anything I had ever heard—so beautiful and powerful that it stirred my soul with joy.

Then, the men guiding me led me further and took me up to the fifth heaven. There, I saw an enormous crowd, too many to count. These were the Grigori, and while they looked like men, they were much larger, even bigger than giants.

Their faces looked tired and sorrowful, and they were completely silent. The entire place felt heavy and lifeless—there was no worship, no joy, and no service to the Lord. Confused, I turned to the men who brought me there and asked, "Why do these beings look so sad and lifeless? Why are they silent, and why is there no worship here?"

They answered, "These are the Grigori. Long ago, they and their leader, Satanail, turned away from the Lord. Because of their rebellion, they were cast into great darkness in the second heaven. Three of them were sent down from God's throne to a place called Ermon. There, at Mount Iermon, they saw the daughters of men, found them beautiful, and took them as wives.

By doing this, they disobeyed God and brought corruption to the earth. They abandoned their duties and went against His will. Their children became giants, men of great size and strength, but their existence only brought wickedness and chaos. Sin and lawlessness spread everywhere because of them.

Because of their actions, God passed a severe judgment on them. Now, they mourn for their fallen brothers, knowing they will face punishment on the great and terrible day of the Lord."

Hearing this, I turned to the Grigori and said, "I have seen what happened to your brothers and the suffering they endure. I prayed for them, but the Lord has decided they must remain trapped under the earth until the heavens and the earth pass away. They will never be set free."

I continued, "Why are you waiting, my brothers? Why do you not serve before the Lord? Why do you not fulfill your duties and give Him the honor He deserves instead of continuing in your rebellion?"

When I finished speaking, the Grigori listened to my words. They arranged themselves into four groups within the heaven. As I stood with the men who guided me, four trumpets sounded together, filling the air with a deep and powerful sound. Then, the Grigori began to sing as one. Their voices were filled with sorrow, but their song was soft and moving. Their mournful song rose up before the Lord, carrying their regret and longing for redemption.

Chapters XVIII. 4XIX. 2

[The taking up of Enoch, i1tto the sixth, Heaven.]

The men guiding me then took me further and brought me to the sixth heaven. There, I saw seven groups of angels, each one glowing with incredible brightness. Their faces shone even brighter than the sun's rays, and their beauty was beyond anything I had ever seen. They all looked the same, with no differences in their appearance, expressions, or clothing. They stood together in perfect unity.

These angels were responsible for studying and organizing the movements of the stars, the phases of the moon, and the paths of the sun. They ensured that balance was kept in the world, controlling both good and bad conditions as they had been commanded. They also arranged teachings and instructions and created beautiful songs and melodies filled with praise and glory.

These angels were archangels, the leaders over all the other angels. They had authority over everything, both in heaven and on earth. Some were in charge of keeping track of the changing seasons and the passing of years. Others watched over the rivers and seas, making sure the waters flowed as they should. There were also angels who oversaw the growth of plants and trees, ensuring that all living creatures received the food they needed.

I also saw angels who were responsible for recording the lives and actions of every person on earth. They carefully wrote down everything before the Lord, making sure that no deed—good or bad—was ever forgotten.

In the middle of these angels, I saw seven phoenixes, seven cherubim, and seven other beings with six wings each. Together, they sang in perfect harmony with one voice. Their song was so beautiful and powerful that words could not describe it. It was a joyful offering to the Lord, rising up to His holy throne as a tribute to His greatness and glory.

Chapters XIX. 3—XX. 3

[Thence Enoch is taken into the seventh. Heaven]

The men guiding me took me even higher, bringing me to the seventh heaven. There, I saw a breathtaking and brilliant light. Surrounding it were powerful archangels, spirits of great strength, rulers, and other mighty beings. I saw cherubim and seraphim, shining thrones, and countless watchful eyes. In front of me stood ten groups of radiant beings, each more dazzling than the last. The sight was so overwhelming that I trembled in fear, unable to fully comprehend what I was seeing.

Sensing my fear, the men with me held me and reassured me, saying, "Do not be afraid, Enoch. Be at peace." Their words comforted me, and I was able to stand among them.

Then, they showed me the Lord from a distance, seated on His magnificent and exalted throne. His presence was beyond words— majestic and awe-inspiring. Around Him, all the heavenly beings were gathered, each standing on one of ten steps, perfectly arranged according to their rank. They bowed before the Lord, showing Him deep reverence and honor.

After paying their respects, they returned to their places. With joy and devotion, they stood in the endless light of His presence. In soft and harmonious voices, they sang praises, glorifying the Lord and serving Him with love and honor in the brilliance of His unending glory.

Chapters XX. 4—XXII. 5

[How the Angels placed Enoch there at the limits of the seventh Heaven and departed from him invisibly.]

They never leave, day or night, but remain before the Lord, carrying out His commands. Surrounding His throne are the cherubim and seraphim, along with six-winged beings who cover His throne with their presence. They sing softly, their voices full of reverence, proclaiming, "Holy, Holy, Holy, Lord God of Hosts! Heaven and earth are filled with Your glory!"

After witnessing these incredible things, the men who had guided me turned to me and said, "Enoch, our task is complete." Then they departed, leaving me alone. Standing at the edge of heaven, fear overtook me. I fell on my face, trembling, and cried out in distress, "What is happening to me?"

But the Lord, in His mercy, sent one of His great archangels—Gabriel. Gabriel spoke gently to me, saying, "Be at peace, Enoch. Do not be afraid. Stand up and follow me, for you will always remain before the Lord." Though his words were meant to calm me, I answered, "O Lord, my spirit is overwhelmed with fear. Please send back the men who brought me here. They were my companions, and with them, I would feel safe to approach You."

Gabriel, however, swiftly lifted me as if I were a leaf carried by the wind. He brought me directly before the Lord. Overcome with awe, I fell on my face and worshiped Him. Then the Lord Himself spoke to me, saying, "Be at peace, Enoch. Do not fear. Stand before

Me, for you will remain in My presence forever."

Then Michael, the chief of the archangels, approached and lifted me up. He presented me before the Lord, and the Lord commanded His heavenly servants, "Let Enoch stand before Me forever." The glorious beings bowed low before the Lord and answered, "Let it be done according to Your word, O Lord."

Then the Lord turned to Michael and said, "Remove Enoch's earthly garments and anoint him with My holy oil. Then clothe him in the robe of My glory." Michael obeyed. He took away my earthly robe and anointed me with oil that shone brighter than the sun. The oil smelled sweeter than the finest myrrh, felt cool like morning dew, and glowed with brilliant light. Then he dressed me in garments that radiated with divine glory.

As I looked at myself, I saw that I had been transformed. I was now like the glorious beings who serve the Lord, and all fear and trembling left me.

Then the Lord called upon another archangel, Vretil, who was known for his wisdom and for recording all of the Lord's works. The Lord said to him, "Bring the books from My storehouses and give Enoch a reed to write. Teach him the knowledge contained in these books."

Vretil obeyed immediately, bringing books that smelled of myrrh and handing me a reed. With patience and wisdom, he prepared to teach me the mysteries written within those sacred texts.

Chapters XXII. 6—XXIII. 4

[Of the writing of Enoch how he wrote about his wonderful Goings and the heavenly Visions, and he himself wrote 366 Book.]

He showed me how the heavens, the earth, and the seas work. He explained the movement of their elements, the rumble of thunder, the paths of the sun and moon, and the way the stars travel and change. He described the cycles of the seasons and years, the passage of days and hours, the way the winds move, the countless angels in heaven, and the beautiful songs they sing in perfect harmony.

He also revealed everything about people—their lives, their different languages, the songs they sing, their knowledge, and the lessons they follow. He shared the melodies of their voices and all the wisdom they need to understand.

For thirty days and thirty nights, Vretil taught me without stopping, speaking the entire time. And for those same thirty days and nights, I wrote without resting, recording everything he told me.

Then, when the time came, Vretil said to me, "You have written everything I have taught you. Now, write about the souls of all people—the souls that have not yet been born and the places that have been prepared for them for eternity. Every soul was created to live forever, even before the world was made."

I obeyed, writing without stopping for another thirty days and nights. By the time I finished, I had written 366 books, carefully recording everything I had seen and learned.

Chapters XXIII. S—XXIV. 5

[Of the great Secrets of God, Which God revealed and told to Enoch, and spoke with him Face to Face.]

The Lord called me and said, "Enoch, sit at My left side with Gabriel." I bowed down in deep respect before Him.

Then the Lord spoke again, saying, "Enoch, everything you see—whether still or moving—was created by Me. Now, I will show you how I made everything from nothing and how the visible world came from the unseen.

Even My angels do not know these secrets. I have not told them how creation began, and they do not understand the endless greatness of My works, which I am revealing to you today.

Before anything existed, I alone moved through the unseen, just as the sun moves across the sky from east to west. But unlike the sun, which has a place to rest, I never stopped, for I was constantly creating. I shaped the foundation of everything and began to bring visible things into being."

Then the Lord continued, "I commanded the deep abyss so that things could emerge from the invisible. From this unseen realm, Adoil appeared before Me—vast and magnificent, glowing with a brilliant red light.

I said to Adoil, 'Break open, and let what is inside you be seen.' He obeyed, and a great light burst forth. I was surrounded by this brilliant light, and from within it, the entire world was revealed, just as I had planned. I saw that it was good.

Then I made a throne for Myself and sat upon it. I commanded the light, 'Rise high and become the foundation for everything above.' The light obeyed, ascending to the highest place. As I sat on My throne, I looked at the light and marveled at how great it was."

Chapters XXV. 1XXVII. 1

[God again calls from the Depths and there came
Arkhas, Tazhis, and one who is very red.]

I called out again into the emptiness and said, "Let something solid and visible come from what cannot be seen." In response, Arkhas appeared—thick, heavy, and glowing with a deep red color.

I told Arkhas, "Separate, and let what comes from you be seen."

When he split apart, a vast and shadowy world was revealed—huge, dark, and endless—bringing the beginning of everything below.

I saw that this was good. I commanded, "Go down and become the base for all that will be beneath." And so it happened. Arkhas sank, became firm, and formed the foundation for everything below. Beyond this darkness, there was nothing more.

Chapters XXVII. 2XXIX. 3

[How God established the Water, and surrounded it with
Light, and established upon it Seven Islands.]

I commanded the light and darkness to be separated and said, "Let there be something thick and solid." And so it appeared. I spread this new substance out, forming water beneath the light, covering the darkness below.

I made the waters firm, shaping the deep places, and surrounded them with light. Then, I created seven layers, making them clear and strong—both smooth and rough—like glass and ice. I set paths for the waters and other elements, guiding them to move in harmony with the seven stars, each within its own part of the sky. I saw that this was good.

I separated light from darkness and divided the waters above from those below. I told the light, "You will be called day," and the darkness, "You will be called night." And so, the first day began with evening and ended with morning.

Next, I made the heavenly layers stronger and gathered the waters below into one place, keeping the waves under control. From these waves, I formed large, solid stones, and from the stones, I shaped dry land, which I named earth. At the center of this land, I created a deep, endless pit.

I brought the sea together in one place and set a boundary for it, saying, "This is where you will stay forever. You will not cross the limits I have set for you." Then, I made the sky firm and placed it above the waters. This was the end of the first day.

When evening and morning passed, the second day began.

For the heavenly beings, I gave them a nature like fire. I looked at a strong, unbreakable stone, and from the brightness of My eye, I gave lightning its powerful glow. I placed fire within the water and water within the fire, making sure that neither would destroy the other. That is why lightning shines brighter than the sun, and soft water can wear down the hardest stone.

From the stone, I brought forth powerful fire, and from this fire, I created countless spiritual beings—ten thousand angels, each armed with weapons of flame and robes of burning light. I commanded them to take their places and follow the purpose I had set for them.

Chapters XXIX. 4XXX. 3

[Here Satanail was hurled from the Heights with his Angels.]

One of the archangels, leading those beneath him, had a thought that could never come true—he wanted to raise his throne above the clouds and have the same power as Me. Because of this, I cast him down from the heights, along with his followers. Now, he wanders endlessly in the air above the abyss.

With that, I finished creating all the heavens, completing the third day. On this day, I commanded the earth to grow tall, fruit-bearing trees, towering mountains, and every kind of plant and seed. I also created Paradise, surrounding it with a protective barrier and placing fiery, armed angels at its entrance to guard it and keep it forever renewed.

When evening passed and morning arrived, it was the fourth day. On this day, I decorated the sky with great lights. In the highest circle, I placed the star Kruno. In the second, I set Aphrodite; in the third, Ares; in the fourth, the Sun; in the fifth, Zeus; in the sixth, Hermes; and in the seventh, the Moon.

I filled the sky below with countless smaller stars, making the Sun shine during the day and the Moon and stars glow at night. I gave the Sun its path through the signs of the Zodiac and set the Moon to follow the same twelve signs. I fixed their names, their purpose, and the timing of their movements—even the sounds of thunder and the precise passing of time.

When evening passed and morning arrived, the fifth day began. On this day, I commanded the sea to be filled with all kinds of fish and winged creatures. I also brought forth crawling creatures, four-legged animals, and everything that moves through the air. Each one was made male and female, given breath and life.

Finally, evening passed and morning arrived, marking the sixth day. On this day, I turned to My Wisdom and directed it to create man, forming him from seven elements.

Chapter XXX. 48

I shaped his body from the earth.
I made his blood from the morning dew.
His eyes came from the light of the sun.
His bones were formed with the strength of stones.
His thoughts were drawn from the speed of angels and the
    drifting clouds.
His veins and hair grew from the grass of the land.
And his spirit came from My own breath and the wind itself.
I gave him seven unique abilities:

His body could hear, his eyes could see, his mind could recognize scents, his veins could feel touch, his blood could taste, his bones could endure, and his thoughts carried both sweetness and wisdom.

I designed man with a perfect balance of what is seen and unseen. His life and death, his shape and spirit, all connected to both realms. Though his creation was small compared to My endless power, it carried a great and deep purpose.

I placed him on the earth as a being like no other, almost like an angel in human form. I gave him honor, strength, and glory.

I made him ruler over the earth, guiding it with My wisdom. Among all that I created, nothing else was like him.

Chapter XXX. 915

I gave him a name inspired by the four directions—East, West, North, and South.

I placed four guiding stars for him and named him Adam.

I gave him free will and showed him two paths—the path of light and the path of darkness. I told him, "This is good, and this is bad," so that his choices would reveal his heart. Through him, his descendants would also show their true nature, whether they loved Me or turned away.

Even though I understood his nature, he did not yet understand himself. This lack of knowledge became his struggle, leading him to make mistakes. Because of his wrongdoing, I declared that death would be the price to pay.

I put him into a deep sleep, and while he slept, I took one of his ribs and created a companion—his wife.

Through her, death entered the world, and I accepted what would become of his descendants. I gave her a name, calling her the mother of all living—Eve.

Chapter XXX. 16, 17

[Goel gives Paradise to Adam, and gives him Knowledge, so as to see the Heavens open, and that he should see the Angels singing a Song if Triumph.]

Adam lived his life on earth, and I created a garden in Eden, in the East, for him. I gave him the responsibility to follow My instructions and care for what I had given him.

I opened the heavens so he could see the angels singing songs of victory. In Paradise, there was always light—no darkness existed there.

But the devil, filled with jealousy, wanted a world of his own because everything on earth was under Adam's rule.

The devil, an evil spirit from the lowest places, became known as Satan after leaving heaven. Before his fall, his name was Satanail.

Though his nature changed, making him different from the angels, he still knew the difference between right and wrong. He fully understood the judgment against him and the sin he had committed.

Out of spite, he plotted against Adam and tricked Eve. But he never directly controlled or touched Adam.

Because of his evil plans, I cursed him for his arrogance and wickedness. However, I did not curse those I had already blessed.

I did not curse man, the earth, or anything I had created. Instead, I cursed the results of man's disobedience and the corruption that came from it.

Chapters XXX. 18XXXIII.1

[On account of the Sin of Adam, God sends him to the Earth,
'From which I took thee,' but He does not wish to destroy him in
the Life to come.]

I told him, "You were made from the earth, and one day, you will return to it. I will not destroy you, but you will go back to where you came from. Then, when I return, I will take you again."

I blessed everything I created, both seen and unseen.

I also blessed the seventh day, the Sabbath, because on that day, I rested from all My work.

Chapter XXXIII. 26

[God shows Enoch the Duration of this World, 7000 Years, and the
eighth Thousand is the End. (There will be) no Tears, no months,
no Weeks, no. Days.]

Then I established the eighth day, making it the first after all My work was done. Let this day represent a time without end—no more counting years, months, weeks, days, or hours, but an everlasting age beyond measure.

Now, Enoch, everything I have told you, all that you have seen in heaven and on earth, and everything you have written—know that I created it all with My wisdom. From the highest places to the lowest, from the beginning to the end, I designed everything. No one advises Me or shares in My work, for I am eternal and not made by anyone. My thoughts never change, My wisdom guides Me, and My word is always true. I see all things, and when I look upon them, they remain as they are. But if I turn away, all things depend on Me to exist.

Listen carefully, Enoch, and understand who is speaking to you.

Take the books you have written, for I will send you back with Samuil and Raguil, who brought you here. Return to the earth and tell your sons everything I have revealed to you—all that you have seen, from the lowest heaven to My throne.

I created all the heavenly beings and powers, and none stand against Me or disobey My will. Everything follows My command and exists under My authority. Give your writings to your sons, and they will read them and understand that I am the Creator of all things. They will know there is no other God besides Me.

Your writings will be passed down through their children, spreading across generations and nations. I will send you, Enoch, with My messenger, the great leader Michael, to keep these writings safe alongside the records of your ancestors—Adam, Seth, Enos, Kainan, Mahalaleel, and your father Jared.

These writings will not be needed until the final age. For this reason, I have assigned two angels, Arinkh and Parinkh, to guard them on earth. They will make sure that the story of what happens to your family is not lost when the great flood comes.

Chapters XXXIII. 7XXXV. 1

[God accu.1es the Idolators; the Workers of Iniquity, such as Sodom, and on this account, He brings the Deluge upon them.]

I know how corrupt people will become. They will refuse to follow the path I set for them or use the gifts I have given. Instead, they will reject My guidance, follow another way, and invest in things that have no value. They will worship false gods and turn away from Me, the one true God.

The earth will be filled with evil, wickedness, and impurity. People will harm one another in terrible ways, committing sins too awful to name. Because of this, I will bring a great flood to wipe out everything, for the world will have become completely corrupt.

But I will save one righteous man from your family line, along with his household, because they will follow My ways. Over time, their descendants will grow into a large nation, though many among them will be consumed by their own selfish desires.

When their time ends, I will reveal to them the books you have written, along with the writings of your ancestors. The keepers of these books on earth will share them with those who are faithful to Me—those who respect My name and do not dishonor it. These people will pass the knowledge to the next generation, and those who read it will bring Me even greater glory.

Now, Enoch, I am giving you thirty more days to stay on earth and teach your family. Gather your sons and relatives before Me so they may hear your words. Let them read and understand that there is no other God but Me. Teach them to obey My commandments always and study the books you have written.

After thirty days, I will send My angels to take you away from the earth and from your sons, just as I have planned.

Chapters XXXV.2 XXXIX. 1

[Here God summons an Angel.]

The angel standing beside me was a sight both breathtaking and terrifying. His appearance was as pure as freshly fallen snow, and his hands felt as cold as ice. The chill from his presence sank deep into my face, overwhelming me, for I could not bear the immense power of the Lord. It was like trying to withstand the heat of a blazing fire or the biting cold of the harshest winter.

Then the Lord said to me, "Enoch, if your face does not grow cold in this place, no human on earth could ever look at it and survive."

Meanwhile, my son Mathusal waited with hope, keeping watch by my bedside day and night, longing for my return. The Lord spoke to the ones who had taken me and commanded, "Return Enoch to the earth and remain with him until the appointed time." That night, they brought me back to my bed, where Mathusal had been keeping faithful watch. When he heard me return, he was filled with fear. I called for my whole household to gather because I had much to tell them.

With sorrow in my heart and tears in my eyes, I spoke to my children, filled with deep sadness.

"Listen carefully, my children, to the words given to me by the Lord. Today, I have been sent to you by Him to tell you what has happened, what is happening now, and what will take place before the day of judgment.

"Pay close attention, for these words are not my own. They come directly from the Lord, who has commanded me to share them with you. I am only a man, like you, but I have seen the face of the Lord. It was like metal glowing red-hot in a fire, sending out sparks that burned everything in their path."

"Look into my eyes, the eyes of a man who has been sent to deliver this message. I have looked into the eyes of the Lord, eyes that shine like the sun and fill the hearts of men with fear. Look at my hands, made of flesh like yours. I have seen the right hand of the Lord, a hand so powerful it stretches across the heavens, bringing help and support."

"My actions are human, just like yours, but I have seen the boundless, perfect form of the Lord, a form without limit or measure. You hear my voice, but I have heard the voice of the Lord. His words thunder like a storm, rolling through the sky with the power of roaring clouds."

"My children, listen to the words of your father. You know how terrifying it is to stand before a ruler on earth, knowing that your life or death depends on their judgment. But how much more fearsome, how much more overwhelming, is it to stand before the face of the Lord of all lords, the Master of heaven and earth? Who among us could ever endure such endless fear and trembling?"

Chapters XXXIX. 2XL. 6

[Enoch instructs fitfully his Children about all 1'hings from the Mouth of the Lo1'(l; how he Mw, and heard and wrote them clown.]

And now, my children, I want you to understand that the Lord has revealed everything to me. My eyes have seen all things, from the very beginning to the end of time. I have written down everything I have witnessed in my books—the vast heavens, their endless space, and the countless beings that fill them. I have studied the paths of the stars and recorded their movements, even though their number is too great to count.

No human has ever seen the full course of the stars, and not even the angels know how many exist. Yet, I have written down each of their names. I have measured the path of the sun and the strength of its rays, tracking its rising and setting throughout every month of the year. I have carefully recorded all its movements and given them their proper names.

I have also studied the moon's orbit, tracking its daily phases and the hidden places it retreats to before rising again. I have followed its journey through time, measuring its path by the hours. I have defined the four seasons and divided them into four great cycles. Within these cycles, I arranged the years, set the months in order, and from the months, counted the days. From the days, I measured the hours.

Beyond the heavens, I have observed everything that moves upon the earth. I have recorded every living creature, every plant that is sown or grows naturally, and all vegetation found in gardens. Every herb, flower, and fragrance has been written down along with its name.

I have studied how clouds form and move, how they gather water and release rain. I have observed how raindrops fall and documented everything I have learned. I have tracked the paths of thunder and lightning, and I was shown the forces that control them. I saw the guardians who hold the keys to these powerful forces, releasing them in measured amounts so they do not destroy the earth.

I have recorded the storehouses of snow and hail and studied the gentle breezes. I saw the keeper of these elements, how he fills the clouds with snow and hail without ever running out. I observed the places where the winds rest and watched their keepers carefully measure and release them, ensuring they do not shake the earth with too much force.

I have measured the entire earth—its tallest mountains, rolling hills, vast fields, and thick forests. I have recorded the stones, rivers, and everything that exists on the land. I measured the height from the earth to the seventh heaven and the depths down to the lowest place of judgment. There, I saw the great abyss open, filled with cries of despair, where souls suffer as they await their final judgment.

I wrote down the names of those being judged, recording the punishments they received and the actions that led to their sentence. Every deed and its judgment have been documented in my writings, preserving the truth of what I have seen.

Chapter XL. 713

[How Enoch wept for the Sins of Adam.]

I saw all the people who came before us, starting with Adam and Eve, and I felt a deep sadness. Tears filled my eyes as I thought about the harm their mistakes had caused. Overwhelmed with sorrow, I cried out, "How unlucky I am, struggling with my own weaknesses and the failures of those before me!"

In my heart, I thought, "The luckiest people are the ones who were never born—or if they were, never did anything wrong against the Lord. They would never have to see this place or suffer its pain."

Chapters XLI. 1XLII. 6

[How Enoch Baw tho8e who keep the Keys, and the Guardians of the Gates of Hades standing by]

I saw the gatekeepers of hell, standing tall like giant serpents. Their faces were dark and empty, like lamps that had gone out. Their fiery eyes burned fiercely, their sharp teeth shone, and they wore nothing on their upper bodies.

I stood before them and said, "I wish I had never seen you or heard of what you do. I wish no one from my family had ever come to this place. The people of my kind have made mistakes during their short time on earth, but now they must suffer forever."

From there, I traveled east to the paradise of Eden, a place of rest prepared for the good and righteous. It is connected to the third heaven but is hidden from this world. At its grand gates, where the sun rises, stand powerful angels surrounded by flames. They sing songs of victory, celebrating endlessly in the presence of the just.

On the final day, Adam and our ancestors will be brought into this paradise. They will enter with joy, like guests invited to a great feast. Together, they will arrive with happiness, share in cheerful conversations, and eagerly wait for the celebration—a feast filled with never-ending light, endless blessings, and a life of joy and laughter.

Then I said, "Listen to me, my children: Happy is the one who respects the Lord's name and serves Him with honesty. Blessed is the one who gives offerings with a sincere heart, who lives fairly, and who dies in righteousness.

Blessed is the one who judges fairly, not for rewards but because they love what is right. In the end, they will receive true justice. Blessed is the one who gives clothes to those in need and feeds the hungry. Blessed is the one who treats orphans and widows fairly and stands up for those who have been wronged.

Blessed is the one who turns away from the temporary and unstable things of this world and chooses the path of goodness, leading to eternal life. Blessed is the one who does good deeds, for they will receive even greater rewards.

Blessed is the one who speaks truthfully and has a kind and loving heart. Blessed is the one who understands the works of the Lord and gives Him praise. The Lord's ways are always right, and while people may do good or bad, each person is known by their actions."

Chapters XLII. 7—XLIII. 2

[Enoch shows his Children how he measured and wrote out the
Judgements of God.]

Listen, my children, to the wisdom I have gathered in my life and the lessons I have reflected on from the Lord. I have written these thoughts down through the seasons, both in winter and summer. I have studied the passage of time, measured the years and hours, and carefully recorded their changes.

Just as one year can stand out more than another, people also differ from one another. Some are respected for their wealth, while others are honored for their wisdom. Some are known for their deep understanding, while others are admired for their clever thinking. A person might be valued for speaking softly, another for having a pure heart. Strength makes some stand out, while beauty makes others shine. Youth brings energy, while sharp thinking brings recognition. Some are praised for their sharp senses, while others are admired for their ability to understand and learn many things.

But let everyone remember this: no one is greater than the one who respects and follows God. That person will be the most honored and will remain strong and righteous forever.

Chapters XLIII. 3XLF. 4

[Enoch instructs his Sons that they should not revile the Person, of men, whether they are great or small.]

God created people with His own hands, shaping them in His own image. He made both the powerful and the weak, and anyone who mocks another person's appearance is insulting God Himself.

If someone becomes angry at another without a good reason, they will face the Lord's great anger. If a person spits on someone in disrespect, they will stand before God's judgment and face its consequences.

Blessed are those who hold no hatred in their hearts. Blessed are those who defend the mistreated, support the accused, lift up the oppressed, and answer the cries of those in need.

For on the day of judgment, every act of fairness—every scale, measurement, and tool of justice—will be tested and will receive its true reward.

Chapters XLVI. 1XLVIII. 1

[God shows that He does not wish Sacrifices from Man, nor Burnt Offerings, but pure and contrite Heart]

Whoever brings their offerings quickly before the Lord will also receive His blessings just as swiftly. God will ensure that justice is done for them. Whoever lights a lamp in His honor will find that their treasures in heaven grow even greater. But God does not actually need bread, light, animals, or any material gifts. These things have no real value to Him. What He truly desires is a heart that is honest and devoted. Through these offerings, He looks at what is inside a person's heart and tests their sincerity.

Think about how a king on earth would react if someone gave him a gift while secretly hiding bad intentions. If the king realized the person was being dishonest, wouldn't he reject the gift in anger and punish them? In the same way, if someone speaks kindly to another but secretly plans to hurt them, their deception will eventually be exposed, leading to shame. If people react this way, imagine how much more God despises and rejects gifts that come from dishonest hearts. He does not accept such offerings but instead turns away from them in anger.

One day, God will send a great light that will reveal everything. Both the good and the bad will be judged. No secret will remain hidden; every thought and intention will be exposed by His truth.

Now, my children, I urge you to keep these lessons close to your hearts. Think carefully about the words I have shared with you, for they do not come from me alone but from the Lord Himself. Hold on to these sacred teachings and study them carefully. In them, you will discover the incredible works of God. While many books have been written throughout history and more will continue to be written, none will reveal God's truth as clearly as these words.

If you follow these teachings and live by them, you will not turn away from God. Remember, there is no other God—nowhere in heaven, on earth, beneath the ground, or in the hidden depths of creation. Only He laid the unseen foundations of the world, stretched out the skies both visible and invisible, and set the earth upon the waters. He holds the waters in place without any solid ground beneath them, shaping everything in its endless beauty and variety.

Who but God can count every speck of dust on the earth, every grain of sand on the shore, every raindrop that falls, or each drop of morning dew? Who can measure the wind or control the land and sea with unbreakable laws? Who shaped the stars from fire, decorated the sky, and placed the sun at its center, giving it light and warmth for all? Only the Lord, the Creator of everything, has done these things and more. His power has no limits, and His wisdom is beyond understanding.

Chapters XLVIII. 2XLIX.1

[Of the course of the Sun throughout the seven Circles.]

The sun moves through the seven layers of the sky, and I have given it 182 positions for the days when its path is shorter and 182 for when its path is longer. In addition, there are two special positions where it rests as it moves between its monthly cycles. Starting in the month of Tsivan, after seventeen days, the sun begins to move downward until the month of Thevan. Then, on the seventeenth day of Thevan, it begins to rise again.

This is how the sun follows its path in the sky. When it comes closer to the earth, the land is filled with joy, bringing an abundance of life and fruit. But when it moves farther away, the earth becomes dull, and trees and plants stop growing. Everything follows a precise and perfect order, set by God's endless wisdom, both in the visible world and beyond what we can see.

From the unseen, He created everything we see, though He Himself remains invisible. That is why, my children, I encourage you to share these writings with your families, your children, and people everywhere. Let those who are wise and respect God treasure these words. Let them value them more than the finest food, reading them with care and devotion.

But those who lack understanding and refuse to think about God will reject these teachings and turn away from them. For such people, judgment will come, and they will face the consequences.

Blessed is the one who accepts these teachings, carries them with faith, and follows them in life. That person will be free on the day of judgment and will stand in the light of truth and righteousness.

Chapters XLIX. 2LI. 2

[Enoch instructs his Sons not to swear either by the Heaven or the Earth; and shows the Promise of God to a Man even in the Womb of his Mother]

I tell you this, my children, with complete honesty. I will not swear by heaven, earth, or anything that God has made, because God Himself has said, "There is no swearing in Me, no injustice—only truth." If people do not have truth in their hearts, they should simply say "yes" when they mean yes and "no" when they mean no.

But I want you to know for certain—there has never been a single person born for whom a place has not already been prepared. Every soul has a purpose, and the time each person spends on this earth has already been set. So do not be misled, my children. Every soul has its own path and destination.

No one who has ever lived can hide from God, and nothing they do is truly secret. He sees everything, and I have written down the actions of every person. Each of us is given only a short time on earth, where we must face challenges and hardships. But during that time, we must never harm those who are vulnerable, like widows and orphans.

So, my children, live your days with patience and humility, and you will receive the gift of eternal life. Endure pain, hardships, and cruel words for the sake of the Lord. If someone wrongs you, do not seek revenge—not against a neighbor or even an enemy. Leave justice to God, for He alone will judge and repay when the time comes. Seeking revenge is not your place.

If any of you share your wealth to help a brother in need, you will be greatly rewarded on the day of judgment. Be generous to orphans, widows, and strangers, for acts of kindness are seen as treasures in the eyes of the Lord.

Chapters LI. 3LIII. 1

[Enoch instructs his Sons, not to hide their Treasures upon Earth, but lids them give Alms to the Needy.]

Help those in need as much as you can. Do not hide your wealth away—use it to support those who are honest but struggling. If you do, trouble will not come upon you when you face hardship. No matter what difficulties or challenges you must go through, endure them for the Lord's sake, and you will receive your reward on the day of judgment.

It is good to go to the house of the Lord in the morning, afternoon, and evening to honor the Creator of everything. Let all living things praise Him, and let every creature, seen and unseen, join in worship.

Blessed is the one who speaks to glorify the Lord and praises Him sincerely from the heart. But cursed is the one who uses their words to insult or harm others. Blessed is the one who lifts up God's name, but cursed is the one who spends their life speaking with anger, swearing, and showing disrespect.

Blessed is the one who appreciates and honors God's works, while cursed is the one who speaks badly about His creation. Blessed is the one who works hard and takes responsibility for their own efforts, but cursed is the one who relies on others without contributing. Blessed is the one who respects and upholds the traditions of their ancestors, while cursed is the one who disregards or destroys them.

Blessed is the one who brings peace and love among people. Cursed is the one who causes conflict and division. Blessed is the person who carries peace in their heart, not just talking about it but truly living it. However, cursed is the one who pretends to seek peace but secretly holds anger and resentment.

All these actions—both good and bad—are recorded, and on the day of judgment, everything will be revealed.

Chapters LIII. 2LVI. 1

[Let us not say that our Father is with God, and will plead for us at the Day of Judgment. For I know that a Father cannot help his Son, nor a Son a Father.]

Now, my children, do not think you can say, "Our father prays before God to free us from sin," because no one can take responsibility for another person's mistakes. Everyone is responsible for their own actions. I have recorded everything a person will do, even before they are born, just as it has been done for all people throughout time.

No one can change or erase what I have written, because God sees everything, even the hidden thoughts of those who do wrong. Nothing is truly secret from Him.

So, my children, listen carefully to my words. Do not ignore my advice only to regret it later and say, "Our father never warned us when we were lost in our foolishness." Pay attention now so that you do not look back with sadness, wishing you had known and done better.

Chapters LVI. 2LVIII. 5

[Enoch admonishes his Son that they should give the Books to Others.]

Let these books I am giving you be a gift of peace. Do not keep them hidden—share them with anyone who wants to learn. Encourage others to understand the incredible works of the Lord, which are beyond human understanding.

My children, my time with you is almost over. The moment is near when I must leave this world and go to heaven. Look, the angels are already here, waiting for God's command to take me. Tomorrow morning, I will go to the highest heavens, where I will live forever. So I urge you to do what is right in the eyes of the Lord and follow His ways.

Hearing this, Methuselah said to his father, "If it pleases you, Father, let me bring you food. Then, bless our homes, your sons, and our entire family so that your blessing may bring honor to your people. After that, you may go as God has commanded."

Enoch replied, "Listen, my son. Since the Lord anointed me with His glory, I no longer need earthly food. My soul has moved beyond the pleasures of this world, and I no longer desire anything from it."

Then Enoch said, "Call your brothers, their families, and the elders of the people so I may speak to them before I go, as the Lord has instructed me." Methuselah quickly gathered his brothers— Regim, Riman, Ukhan, Khermion, and Gaida—along with the elders of the people. They all came before Enoch, who blessed them and began to speak.

"My sons, listen to me. When the Lord first came to the earth for Adam's sake, He visited all of His creation. He gathered every animal, every creeping thing, and every bird in the sky and brought them before our father Adam. Adam named every living thing on earth, and the Lord made him ruler over all, placing everything under his care and commanding them to obey him. This is how God created man as the master of His creation.

But the Lord will not judge an animal because of a man's actions. Instead, He will judge people for how they have treated the animals. Just as there is a special place for every human soul, there is also a place for the souls of all creatures. Not a single soul that God has made will be lost before the great day of judgment. On that day, every animal will bear witness against the people who treated them unfairly, for the Lord sees everything and will judge with perfect justice."

Chapters LVIII. 6LIX. 4

[Enoch teaches all his Sons why they must not touch the Flesh of Cattle, because of what comes from it.]

Whoever treats animals cruelly or unfairly is also harming their own soul. When a person offers a clean animal as a sacrifice, they do so to protect their soul, recognizing that the life they take is meant for both nourishment and spiritual purpose. When done properly, this act is considered righteous and brings protection.

But if someone kills an animal carelessly or without respect, they hurt their own soul and commit a sin against themselves. It is a serious wrongdoing to harm any living creature without reason, as it reflects a heart filled with selfishness and cruelty. If a person secretly hurts an animal, it is an evil act that stains their soul, showing a lack of kindness and honesty.

Just as it is wrong to harm animals, it is even worse to harm another human—whether by physical harm or through wicked intentions. If someone causes suffering to another person's soul, they also bring suffering upon themselves, leaving no hope for forgiveness. Taking another person's life not only destroys them but also ruins the soul and body of the one who committed the act, cutting them off from redemption forever.

Anyone who sets a trap for another person will eventually fall into it themselves because deceit does not go unnoticed by God. A person who attacks their neighbor—whether with weapons or with harmful words—will have to face judgment and will not escape punishment. Speaking unfairly or acting unjustly toward someone else is a serious offense that takes away any claim to righteousness.

My children, guard your hearts against all forms of wrongdoing, for these are the things the Lord rejects. Just as you ask God for mercy, you should also show kindness and compassion to every living being. Let your actions reflect the goodness you hope to receive. Help those in need, especially the poor, and give generously from the work of your hands. In the world to come, nothing will be hidden—God sees everything, both good and bad. So live with fairness, kindness, and truth.

Chapters LIX. 5LXII. 1

There are many places prepared for people after this life, each one suited to their actions—good for those who lived righteously and bad for those who lived in evil. These places are endless in number, representing the eternal future of every person. Blessed are those who are worthy to enter the homes of the righteous, where they will live in peace and everlasting joy. But those who are sent to the homes of the wicked will find no rest, no relief, and no hope of escape.

Listen carefully, my children, both young and old. When a person thinks good thoughts and offers gifts to the Lord from their own hard work, their offering must come from what they have earned through honest effort. If someone presents a gift that they did not work for, the Lord will reject it, and it will bring them no benefit. Likewise, if someone works but complains in their heart and gives unwillingly, their gift will not be accepted, and they will gain nothing from it.

When offering gifts to the Lord, do so with faith and sincerity. Blessed is the person who brings their offering with patience, humility, and devotion, for this act can help atone for their sins. But do not waste time with empty talk or delay doing what must be done, for missing the right moment leads to loss. After death, there is no chance to make up for missed opportunities—what is lost on earth is lost forever. Doing things at the wrong time offends both people and God because it shows disrespect for His divine order.

When you see someone in need, do not look down on them. Instead, help them with kindness and a sincere heart. When a person gives clothing to the poor or feeds the hungry, they earn a reward from the Lord. But if they give while complaining or with a reluctant heart, they commit two wrongs—they take away the value of their gift and lose their own reward. Those who receive help must also be careful, for if a poor person takes what is given with pride or ungratefulness, they waste the lesson of their hardship and miss the chance to be blessed in return.

The Lord despises arrogance and lies. Every proud or hateful word, every dishonest act covered in deceit, is offensive to Him. Such wickedness is like a sharp sword that cuts through truth, and it will be thrown into the fire to burn forever. So, my children, live with humility, honesty, and faith, so that your actions will be pleasing to the Lord and bring you everlasting reward.

Chapters LXII. 2LXV.2

[How the Lord call8 Enoch: the People take Counsel to go to kiss him in the Place called Achuzan.]

When Enoch spoke these words to his sons and the elders, news of how the Lord had called him spread quickly. People from near and far heard about it and said to one another, "Let us go and see Enoch and honor him!"

Around two thousand people gathered at a place called Achuzan, where Enoch and his sons were staying. The elders and leaders of the people approached Enoch with great respect. They bowed before him, kissed him, and said, "Enoch, our father, may the Lord, the eternal King, bless you! Today, we ask that you bless your sons and all of us here so that we may be honored in your presence."

They continued, "You, Enoch, are forever glorified before the Lord. God has chosen you above all people on earth and made you the scribe of His creation, recording everything that is seen and unseen. You stand against the sins of men and protect your family."

Enoch listened to them and then spoke these words to his sons and to all who had gathered:

"Listen carefully, my children. Before anything existed, before any creature was made, the Lord created everything, both seen and unseen. When the right time came, He made man in His own image and likeness. He gave him eyes to see, ears to hear, a heart to understand, and the ability to think, make choices, and seek wisdom.

The Lord designed the world with mankind in mind. He created everything for man's sake, setting specific times for all things. From these times, He made years, from years He formed months, from months He shaped days, and from days He set seven in a cycle. Within these seven, He divided the hours into smaller parts so that man could understand the seasons and keep track of time—years, months, and hours. By knowing this, people could reflect on their lives from beginning to end, recognize their sins, and remember both their good and bad deeds.

For nothing is hidden from the Lord's sight. Each person must be aware of their actions and avoid breaking His commandments. Keep my writings safe and pass them down to future generations, for they contain wisdom and guidance for all.

When the time comes for all things, both visible and invisible, to reach their end—when the creation that the Lord Himself has made is completed—then all humanity will stand before Him in the great judgment. On that day, time itself will come to an end, and…"

Chapters LXV.2—LXVI.1

There will come a time when days, months, years, and hours will no longer exist. Time itself will stop being measured. Instead, there will be one eternal age where all those who have lived righteously and escaped God's great judgment will be gathered together to live forever. These good and faithful souls will exist in endless joy and unity, with no end to their happiness.

In this eternal life, there will be no hard work, no sickness, no sadness, no fear, no hunger, and no darkness. There will never be another night. A strong and unbreakable wall will surround them, keeping them safe. They will live in a bright and perfect paradise, where everything that can decay or be destroyed will be gone forever. This paradise will be their home forever, filled with the light of the Lord, free from pain and suffering.

As Enoch spoke to his sons and the elders, he reminded them to live with deep respect for God and to avoid anything that goes against His ways. He warned them to protect their souls from all forms of wrongdoing, for the Lord despises evil. "Serve only the Lord," he said. "Do not worship idols or anything made by human hands. Worship the Creator, who made the heavens, the earth, and everything in them. God is everywhere—in the skies, on the earth, and even in the deepest parts of the sea. Nothing is hidden from Him. He sees all that we do."

Enoch encouraged them to live with patience, humility, and love. He urged them to endure suffering, insults, and temptations, knowing that these struggles were temporary and that their reward would last forever. "Blessed are those who escape the great judgment," he said. "They will shine seven times brighter than the sun, for they have been set apart as righteous."

He reminded them of the order of creation—how God separated light from darkness, created paradise, and prepared the fires of judgment. He recorded all these things so they could read and understand them. His writings were meant to guide them, helping them stay faithful to God's commandments.

After Enoch finished speaking, a great darkness covered the land, like a heavy cloud surrounding the people. Suddenly, angels appeared and carried Enoch up to the highest heaven, where the Lord welcomed him into His presence. As Enoch rose, the darkness disappeared, and light returned to the earth. The people who witnessed this amazing event did not fully understand what had happened, but they praised God and went home, telling others what they had seen.

Enoch was born on the sixth day of the month of Tsivan, and he lived for 365 years. On the exact day and hour of his birth, he was taken up to heaven, completing his life on earth. Before leaving, Enoch spent thirty days writing about all of God's creations, producing 366 books. He left these writings with his sons as a lasting gift of divine wisdom.

After Enoch was taken to heaven, Methuselah and his brothers built an altar at Achuzan, the place where Enoch had ascended. They made sacrifices to the Lord and invited all the elders and people to join in a great celebration. The people brought gifts to Enoch's sons, and for three days, they rejoiced, praising God for the incredible sign He had shown through Enoch, a man who had received an extraordinary blessing from the Lord.

This celebration and the story of Enoch's life were passed down through generations as a reminder of God's greatness and mercy. It served as a lesson for all to remain faithful to the Lord, to love and serve Him, and to pass His commandments from one generation to the next. Amen.

Enoch III
The Hebrew Book of Enoch
(3 Enoch)

Chapter I

Rabbi Ishmael describes his journey into the heavens, where he saw the divine vision of the Merkaba, God's heavenly chariot. As he passed through each level, each one more incredible than the last, he finally reached the entrance to the seventh and highest realm. There, he stood in prayer before the Holy One, looking up at the brilliant light above. In his heart, he called upon the merit of Aaron, the son of Amram—a man known for spreading peace and kindness—who had been given the honor of priesthood directly by God on Mount Sinai.

Rabbi Ishmael prayed with deep emotion, asking that Aaron's righteousness protect him from Qafsiel, the ruling angel, and the other powerful beings guarding the entrance, so they would not harm him or stop him from entering. The Holy One heard his plea and sent Metatron, the Prince of the Presence, a great and exalted angel who serves closest to God's glory.

Metatron, shining with brilliant light, spread his wings in obedience and came down to meet Rabbi Ishmael. Taking his hand with strength yet gentleness, Metatron reassured him in the presence of the other heavenly beings, saying, "Come in peace before the great and mighty King, and behold the splendor of the Merkaba." With that, Rabbi Ishmael was led into the seventh Hall and brought before the dwelling place of the Divine Presence, where he stood before the

Holy One and witnessed the incredible vision of the Merkaba.

The heavenly rulers of the Merkaba and the fiery Seraphim looked upon him, their radiant glow and piercing gaze overwhelming him with fear and awe. Their presence was so powerful that he trembled and collapsed, unable to withstand their brilliance. His strength faded as he was overcome by the splendor of their faces and the overwhelming glory surrounding him.

Seeing this, the Holy One rebuked the Seraphim, Kerubim, and Ophannim, saying, "My servants, my Seraphim, my Kerubim, and my Ophannim, turn your eyes away from Ishmael, my son, my friend, my beloved one, and my glory, so that he will not tremble in fear before you."

At God's command, Metatron, the ever-faithful Prince of the Presence, stepped forward and restored Rabbi Ishmael's spirit. He helped him stand, but even then, Rabbi Ishmael was too weak to speak a single word of praise before the Throne of Glory. The moment was too powerful, and he remained silent in awe. Only after an hour had passed did he finally regain the strength to lift his voice in worship.

Then, the Holy One opened the heavenly gates before him. But these were not ordinary gates—they were the gates of Divine Presence, Peace, Wisdom, Strength, Power, Speech, Song, Holiness, and Praise. As these sacred gates opened, Rabbi Ishmael's eyes were filled with light, and his heart overflowed with praise. He lifted his voice and sang psalms, songs, and praises, pouring out words of thanksgiving, glory, and worship.

As Rabbi Ishmael lifted his voice in a song of devotion and awe before the Holy One, the heavenly Chayyoth, both above and below the Throne of Glory, joined in perfect harmony. Their voices echoed with powerful cries of "HOLY" and "BLESSED BE THE GLORY

OF YHWH FROM HIS PLACE!" Together, their praises rose high, filling the heavens as a tribute to the endless majesty and glory of the Divine King.

Chapter II

Rabbi Ishmael described an incredible moment when the highest angels—the mighty eagles of the Merkaba, the fiery Ophannim, and the blazing Seraphim—began to question his presence. Their voices, filled with authority and devotion to God, called out to Metatron, the exalted Prince of the Presence. Their words carried both curiosity and amazement.

"Young one," they said, addressing Metatron with a title that reflected his strength and closeness to God. "Why do you allow a human, someone born on earth, to enter this holy place and witness the sacred vision of the Merkaba? Where is he from? What tribe does he belong to? What makes him worthy of such an honor?"

Metatron, always faithful to his role, answered with respect and confidence. "He is from the nation of Israel, the people chosen by the Holy One out of all seventy nations. This nation was set apart to carry His name and follow His commandments. He comes from the tribe of Levi, the tribe dedicated to God's service. He is a descendant of Aaron, whom the Holy One Himself chose and honored with the crown of priesthood at Mount Sinai."

When the angels heard this, they accepted his answer. Their awe turned into recognition, and they understood the privilege that had been given to Rabbi Ishmael. Their voices, now filled with praise, spoke of the holiness of Israel.

"Indeed," they said, "this man is worthy to see the Merkaba. Blessed is the nation that has him among their people."

Their words of joy and reverence echoed through the heavenly realms as they declared, "Happy is the people who are given such a blessing!" Their proclamation was not just about Rabbi Ishmael's honor—it was a reminder of the special connection between God and His people, a bond that sets apart those chosen to serve Him.

Chapter III

Metatron has seventy names, but God calls him "Youth."

Rabbi Ishmael said:

At that moment, I asked Metatron, the angel and Prince of the Presence, "What is your name?"

He answered, "I have seventy names, each one tied to the seventy languages of the world. But all of them are connected to my main name, Metatron, the angel of the Presence. However, my King calls me 'Youth' (Na'ar)."

Chapter IV

Metatron is the same as Enoch, who was taken to heaven before the Great Flood.

Rabbi Ishmael said:

I asked Metatron, "Why do you share a name with the Creator and have seventy different names? You are greater than all the heavenly rulers, higher than all the angels, and more beloved than any of God's servants. You have more power, authority, and glory than all the mighty ones. So why, in the highest heavens, are you still called 'Youth'?"

He answered, "I am Enoch, the son of Jared. When the people of the Flood turned away from God and lived in corruption, they rejected Him, saying, 'Leave us alone; we want nothing to do with Your ways' (Job 21:14). At that time, the Holy One took me from the world. He brought me to the heavens so that I could serve as a witness against them for all future generations. This way, no one could ever claim that God was unjust.

But why was everyone destroyed? What had their wives, children, or animals done? Why were their horses, mules, cattle, and all their possessions—even the birds—wiped out by the Flood? If the people had sinned, what wrong did the children commit? What could the animals and birds have done to deserve such destruction? How could anyone say it was fair that the innocent were punished along with the wicked?

Because of these questions, the Holy One took me up while the people were still alive, allowing them to witness my ascension. He made me a testimony to His justice. He then appointed me as a prince and leader among the ministering angels.

At that moment, three powerful angels—Uzza, Azza, and Azazel—stepped forward and accused me before the Holy One. They said, "Didn't the First Ones—the ancient angels—warn You not to create humans?"

The Holy One replied, "I created them, and I will take care of them. I will carry them, and I will save them" (Isaiah 46:4).

When the angels saw me, they protested, "Lord of the Universe! Why is this human allowed to rise to the heavens? Is he not a descendant of those who were destroyed in the Flood? Why has he been brought here?"

The Holy One responded, "Who are you to question Me? I find more joy in this one than in all of you. Therefore, he will be a prince and a ruler over you in the high heavens."

Immediately, the angels accepted God's decision. They came to me, bowed, and said, "Blessed are you, and blessed is your father, for your Creator has given you great honor."

And because I am still young compared to them in days, months, and years, they call me 'Youth' (Na'ar)."

Chapter V

The idolatry of Enosh's generation caused God to remove His presence from the earth.

Rabbi Ishmael said:

Metatron, the Prince of the Presence, explained to me:

From the day that the Holy One removed Adam from the Garden of Eden, His presence, the Shekina, remained on a Kerub beneath the Tree of Life. During that time, angels would descend from heaven in organized groups, traveling through the skies in great numbers to carry out His will across the world.

Adam and his descendants stood outside the gates of the Garden, gazing in awe at the brilliant light of the Shekina. Its radiance spread across the entire world, shining 3,000 times brighter than the sun. Anyone who stood in its light lived without suffering—there were no flies, no gnats, no sickness, no pain, and no harm from demons.

Whenever the Holy One moved—from the Garden to Eden, from Eden back to the Garden, then to the sky, and back again—His presence remained visible to all without causing harm. This divine light stayed on earth until the time of Enosh's generation, when people turned away from God and began worshiping idols.

What did the people of Enosh's time do? They traveled across the earth, gathering silver, gold, gems, and pearls. They built massive piles of treasure and used them to carve gigantic idols, each one as large as

1,000 parasangs. Around these idols, they placed the sun, moon, stars, and planets, believing these celestial forces would serve their idols just as they served the Holy One. Their actions reflected what is written in 1 Kings 22:19: "And all the host of heaven was standing by Him on His right hand and on His left."

But how did they accomplish such a thing? They could not have done this without the help of the fallen angels Azza, Uzza, and Azziel, who taught them the forbidden secrets of magic. Using these dark arts, they learned to control and manipulate heavenly forces to serve their idols.

At that time, the angels who served the Holy One brought their concerns before Him, saying, "Master of the World, why do You still care for humans? As it is written in Psalms 8:4, 'What is man (Enosh) that You are mindful of him?' It does not say 'What is Adam,' but 'What is Enosh,' for he has become the leader of idol worshippers.

Why have You left the highest heavens—the glorious realm where Your exalted Name is praised, the place of Your majestic and elevated Throne in Araboth? The heavens of Araboth, the highest of the heavens, are filled with Your splendor, might, and greatness. Your Throne there is lifted above all things.

And yet, You have come down to live among the children of men, who worship idols and compare You to their false gods. Now, You are on earth, but so are their lifeless idols. Why do You remain among people who have turned their backs on You?"

Immediately, the Holy One removed His Shekina from the earth and withdrew His presence from among them.

At that moment, the angels of heaven, along with the hosts and armies of Araboth—countless in number—gathered around the Shekina. They held trumpets and horns in their hands, forming a great procession as they lifted their voices in songs of praise. Surrounded by their music and worship, the Shekina rose to the highest heavens, just as it is written in Psalms 47:3:

"God has gone up with a shout, the Lord with the sound of a trumpet."

Chapter VI

Enoch Ascends to Heaven with the Shekina, and the Angels Question God

Rabbi Ishmael said:

Metatron, the Angel and Prince of the Presence, explained to me:

When the Holy One decided to bring me up to heaven, He first sent Metatron to carry out His command. In front of everyone around me, Metatron appeared and took me away. He carried me in a brilliant blaze of fire, riding on a chariot of flames pulled by fiery horses—servants of divine glory. Surrounded by a glowing light, I rose up and ascended together with the Shekina to the highest heavens.

As soon as I arrived, the holy angels—Chayyoth, Ophannim, Seraphim, Kerubim, and the Wheels of the Merkaba (the Galgallim)—along with the ministers of the fiery presence, became aware of me. They sensed my approach from an incredible distance—36,000 myriads of parasangs away. Smelling my essence, they were astonished and cried out in disbelief:

"What is this scent of a human? What is this trace of a mortal, formed from a tiny drop of flesh, daring to rise to the highest heavens? How can someone born of the earth enter this place and stand among those who are made of fire?"

Still amazed, they continued, saying:

"How can a being of flesh and blood reach this realm? How can a human, created from dust, stand among those who dwell in divine fire?"

Hearing their protests, the Holy One answered them:

"My servants, my heavenly hosts—my Kerubim, my Ophannim, my Seraphim—do not be troubled or upset by this! Listen carefully and understand. Nearly all the people on earth have turned away from Me. They have rejected My kingdom, abandoned My ways, and chosen to worship idols and false gods. Because of their actions, I have removed My Shekina from the earth and lifted it up to the heavens, far from them.

But this one is different. He is special, chosen, and precious among all who live on earth. He is unique, set apart by his faith, unwavering in righteousness, and pure in his actions. His devotion is greater than anyone else's, and he is worthy of this honor. I have taken him from the world as an offering—a soul of great value, chosen from all beneath the heavens to serve in My presence."

Chapter VII

Enoch Is Lifted to the Throne, the Merkaba, and the Angelic Hosts

Rabbi Ishmael said:

Metatron, the Angel and Prince of the Presence, explained to me:

When the Holy One decided to take me away from the generation of the Flood, He lifted me up on the wings of the Shekina's divine wind. Carried by this sacred force, I rose to the highest heavens, beyond anything that can be understood on earth. He brought me into the magnificent palaces of Araboth Raqia', a realm of incredible beauty and greatness.

There, I saw the glorious Throne of the Shekina, shining with a brilliance beyond words. Around it stood the great Merkaba, the divine chariot, surrounded by countless heavenly beings. I saw the troops of anger, fierce in their power, and the armies of judgment, ready to carry out God's will. Encircling the Throne were the fiery Shin'nim, beings of intense light, and the blazing Kerubim, whose radiance was beyond understanding.

The burning Ophannim, wheels of divine fire, moved with endless energy, while the flashing Chashmallim sent out waves of glowing light and mystery. The Seraphim, creatures of pure lightning and flame, hovered nearby, their presence both overwhelming and humbling.

In the middle of this vast, heavenly assembly, the Holy One gave me a special and sacred role. He placed me before the Throne of Glory to serve and stand in His presence every day, witnessing the greatness and beauty that fill the highest heavens.

Chapter VIII

The gates of the treasuries of heaven opened to Metatron

Rabbi Ishmael said:

Metatron, the Prince of the Presence, told me:

Before the Holy One appointed me to serve at the Throne of Glory, He opened for me three hundred thousand gates of wisdom, understanding, kindness, love, humility, mercy, Torah, and reverence for heaven. Each gate unlocked a deeper level of knowledge and virtue, preparing me for my role.

At that moment, the Holy One increased my wisdom, adding layer upon layer of understanding. He deepened my awareness, sharpening my ability to see the finest details of divine truth. He filled me with knowledge upon knowledge, expanding my ability to comprehend His ways. He poured mercy upon mercy into me, strengthening my compassion for all creation. He gave me instruction upon instruction, enhancing my ability to teach and guide with absolute clarity.

He multiplied my love, making my heart overflow with kindness, and filled me with goodness upon goodness, creating an endless well of virtue within me. He clothed me in humility upon humility, grounding my spirit in true meekness. He strengthened me with power upon power, increasing my abilities beyond understanding. He gave me might upon might, allowing me to stand firm and unwavering in my tasks. He filled me with light upon light, making my brilliance shine even brighter than before.

He enhanced my beauty, increasing it until I reflected the splendor of His presence. He covered me in glory upon glory, making me shine with the radiance of His greatness. With all these gifts, I was honored and blessed, receiving qualities greater than any of the children of heaven. He elevated me above them all, granting me virtues and wisdom beyond what any other heavenly being had ever received.

Chapter IX

Enoch Is Blessed and Transformed with Angelic Features

Rabbi Ishmael said:

Metatron, the Prince of the Presence, explained to me:

After everything that had happened, the Holy One placed His hand upon me and gave me fifty-three unique blessings.

He then lifted me up and expanded my size until I stretched across the entire world, both in length and width.

He caused twenty-two wings to grow on me, with thirty-six wings on each side. Each wing was as vast as the whole world itself.

He gave me three hundred sixty-three eyes, and each one shined as brightly as the great light in the heavens.

He adorned me with unmatched splendor, brilliance, radiance, and beauty, filling me with the light of the entire universe. There was no form of majesty or glory that He did not place upon me.

Chapter X

God placed Metatron on a throne at the entrance of the highest hall and sent a messenger to announce his new role. Metatron was now God's representative, ruling over all the heavenly beings and the leaders of different realms—except for eight powerful princes who carried the sacred name of their King.

Rabbi Ishmael said: Metatron, the Prince of the Presence, explained to me:

"The Holy One, blessed be He, did all of this for me. He created a special throne for me, designed to look like the glorious Throne of God. He covered it with a curtain full of light, beauty, kindness, and mercy, shining just like the one before God's own throne. This curtain was decorated with all the lights of the universe, glowing in their full brilliance.

He placed this throne at the entrance of the Seventh Hall and seated me upon it. Then a messenger traveled through the heavens, declaring:

'This is Metatron, my servant. I have made him a prince and a ruler over all the leaders of my kingdoms and the heavenly beings—except for the eight great princes who bear the sacred name of their King.

From now on, any angel or prince who wishes to bring a matter before me must first go to him. They will speak to him, and he will represent them before me. Whatever command he gives in my name must be followed.

I have placed him under the care of the Prince of Wisdom and the Prince of Understanding, who will teach him the mysteries of heaven and earth, as well as the knowledge of this world and the next.

I have also put him in charge of all the treasuries in the heavenly palaces and all the stores of life that I possess in the highest heavens.'"

Chapter XI

God reveals all hidden knowledge to Metatron.

Rabbi Ishmael said: Metatron, the angel and Prince of the Presence, explained to me:

"From that moment on, the Holy One, blessed be He, showed me every mystery of the Torah and all the deepest secrets of wisdom. He let me understand the true depths of the Law, the thoughts of every living being, and the hidden truths of the universe. Every secret of Creation was made clear to me, just as they are fully known to the Creator Himself.

I carefully observed the mysteries of the universe and the wonders hidden within them. Before anyone even had a secret thought, I already knew it. Before something was created, I had already seen it.

There was nothing in the heavens above or in the depths below that was beyond my knowledge. Even before a person formed an idea, I understood their thoughts. Nothing in the highest places or the lowest depths was hidden from me."

Chapter XII

God gives Metatron a robe of glory, crowns him, and calls him "the Lesser VHWH."

Rabbi Ishmael said: Metatron, the Prince of the Presence, explained to me:

"Because of the deep love the Holy One, blessed be He, had for me—more than for any of the other heavenly beings—He made me a special garment of glory. This robe was covered in dazzling light, and He dressed me in it.

He also created a second robe for me, one of honor, decorated with beauty, brilliance, and majesty. Then, He placed it on me as well.

After that, He made a royal crown just for me. It was set with forty-nine precious stones, each glowing as brightly as the sun.

The light from this crown shone in every direction, spreading across the highest heavens, through all seven levels, and to the farthest corners of the world. Then, He placed it upon my head.

Finally, in front of all the heavenly beings, He called me 'The Lesser VHWH,' as it is written in Exodus 23:21: 'For My Name is in him.'"

Chapter XIII

God writes the letters of creation on Metatron's crown with a flaming pen.

Rabbi Ishmael said: Metatron, the angel and Prince of the Presence, explained to me:

"Because of the great love and kindness the Holy One, blessed be He, had for me—more than for any other heavenly being—He used His own finger to write on the crown placed upon my head. With a pen of fire, He engraved the sacred letters by which the heavens and the earth were created.

These same letters brought the seas, rivers, mountains, and hills into existence. They shaped the planets, stars, and all the forces of nature. The winds, lightning, earthquakes, and thunder came from them, as did snow, hail, and storms. Every element of the world and the entire structure of Creation was formed by these letters.

Each letter on my crown glowed with an unending light. At times, they flashed like lightning. Other times, they burned like torches or flickered like flames. Their brightness was as powerful as the sun, the moon, and the stars, shining across the heavens.

When the Holy One, blessed be He, placed this crown upon my head, all the rulers of the heavenly realms trembled before me. The highest princes and the strongest angels—those greater than all the others who stand before God's Throne—shook with fear.

Even Sammael, the Prince of the Accusers, the mightiest of all the rulers of the heavens, was filled with dread when he saw me.

The angels who govern the forces of nature—the angel of fire, the angel of hail, the angel of wind, the angel of lightning, the angel of wrath, the angel of thunder, the angel of snow, the angel of rain, the angel of the day, the angel of the night, the angel of the sun, the angel of the moon, the angel of the planets, and the angel of the stars—all of them, powerful in their own right, trembled at the sight of me.

These rulers of the world have names:

- Gabriel, ruler of fire
- Baradiel, ruler of hail
- Ruchiel, ruler of the wind
- Baragiel, ruler of lightning
- Za'amiel, ruler of wrath
- Ziqiel, ruler of sparks

- Zi'iel, ruler of disturbances
- Za'aphiel, ruler of storms
- Ra'amiel, ruler of thunder
- Ra'ashiel, ruler of earthquakes
- Shafgiel, ruler of snow
- Matariel, ruler of rain
- Shimshiel, ruler of the day
- Lailiel, ruler of the night
- Galgalliel, ruler of the sun
- 'Ophanniel, ruler of the moon
- Kohbiel, ruler of the planets
- Rahatiel, ruler of the stars

When all these powerful beings saw me, they fell to the ground, unable to look at me. The dazzling light that shone from the crown on my head was so overwhelming that they were struck with awe, unable to lift their eyes to meet mine."

Chapter XIV

Metatron Transformed into Fire

Rabbi Ishmael said: Metatron, the angel and Prince of the Presence, explained to me:

"When the Holy One, blessed be He, chose me to serve at His Throne of Glory, to assist with the divine chariot and the presence of His majesty, my very being was transformed. My body turned into flames of fire, my muscles became blazing sparks, and my bones glowed like burning coals. The light from my eyelids flashed like lightning, my eyes burned like fiery embers, and my hair became flames. Every part of me turned into wings of fire, and my whole form radiated with intense heat.

On my right, streams of fire flowed endlessly, and on my left, burning flames erupted. Stormwinds and tempests surrounded me, while the sound of roaring thunder and shaking earth echoed before and behind me."

Rabbi Ishmael said: Metatron, the highest of all princes, stands before the One greater than all other powers. He moves beneath the Throne of Glory and dwells in a magnificent home of light above. From there, he gathers the fire of deafness and places it into the ears of the heavenly creatures, so they cannot hear the overwhelming voice of God's word.

When Moses climbed to the heights, he fasted for forty days and nights until the secret places of divine energy were revealed to him. He saw the purest, deepest heart within the heart of the Lion, shining as white as its very core. Around him stood countless angels, their presence filled with blazing fire, and they longed to consume him. But Moses prayed—first for the people of Israel and then for himself.

The One who sits upon the divine chariot opened the windows above the cherubim. A vast group of 1,800 angelic beings, along with Metatron, the Prince of the Presence, came out to meet Moses. They gathered the prayers of Israel, shaped them into a crown, and placed it upon the head of the Holy One, blessed be He.

Then they proclaimed, "Hear, O Israel: the Lord our God, the Lord is One." Their faces shone with joy, and the Divine Presence radiated with brilliant light. The Shekina rejoiced, and the angels asked Metatron, "What is this great honor? Who is it for?" The answer came: "It is for the Glorious Lord of Israel."

They declared again, "Hear, O Israel: the Lord our God is One, the Eternal King who lives forever."

At that moment, Akatriel Yah Yehod Sebaoth spoke to Metatron and commanded, "Let no prayer from Moses go unanswered. Listen to his requests and grant them, no matter how big or small."

Then Metatron turned to Moses and said, "Son of Amram, do not be afraid. God is pleased with you! Ask for whatever you desire from His Glory. Your face shines with a light that stretches across the world."

But Moses hesitated. "I fear that I might bring guilt upon myself," he said.

Metatron reassured him, "Take hold of the sacred letters of the oath. They are unbreakable and guarantee that the covenant will never be broken."

Chapter XV

Metatron Divested of His Privilege of Presiding on a Throne of His Own on Account of Acher's Misunderstanding, Thinking Him a Second Divine Power

Rabbi Ishmael said: Metatron, the angel and Prince of the Presence, explained to me:

"At first, I sat on a great throne at the entrance of the Seventh Hall. From there, I carried out judgments for the heavenly beings, ruling over the divine hosts under the authority of the Holy One, blessed be He. I was given the power to grant greatness, kingship, honor, rulership, and glory to the princes of the heavenly realms. While overseeing the Celestial Court, I remained seated, while the princes of the kingdoms stood before me—some to my right and others to my left—all by the command of the Holy One.

But when Acher entered and saw the vision of the divine chariot, he became overwhelmed with fear. His soul trembled, and he nearly lost himself, overcome by terror and awe. He saw me seated on a throne like a king, with countless angels standing around me as attendants. The crowned rulers of the heavens stood in my presence, and the sight filled him with confusion and dread.

In his shock, he spoke aloud and said, 'Surely, there must be two divine powers in heaven!'

Immediately, a divine voice rang out from heaven, from the presence of the Shekina, declaring: 'Return, O wayward children—except for Acher!'

Then Aniyel, a mighty and honored prince, a being of great majesty and power, was sent on a mission from the Holy One, blessed be He. He struck me sixty times with lashes of fire and commanded me to rise to my feet."

Chapter XVI

The Princes of the Seven Heavens, of the Sun, Moon, Planets, and Constellations and Their Hosts of Angels

Rabbi Ishmael said: Metatron, the angel and Prince of the Presence, explained to me:

"There are seven great princes, each of them magnificent, respected, and full of wonder. They have been placed in charge of the seven heavens. Their names are Mikael, Gabriel, Shatqiel, Shachaqiel, Bakariel, Badariel, and Pachriel.

Each of these princes rules over a different heaven and leads a vast host of 496,000 groups of ministering angels.

- Mikael, the highest prince, governs the seventh and highest heaven, Araboth.
- Gabriel, leader of the heavenly armies, rules over the sixth heaven, Maban.
- Shatqiel, another powerful prince, is in charge of the fifth heaven, Ma'an.
- Shachaqiel oversees the fourth heaven, Ja'uf.

- Badariel commands the third heaven, Shejaqim.
- Barakiel rules over the second heaven, Raqia'.
- Pachriel governs the first heaven, Wilan, which is within Shamayim.

Below them is Algalliel, the prince responsible for the movement of the sun. He is accompanied by 96 great and honored angels who guide the sun's path through the heavens of Raqia'.

Beneath them is Ophanniel, the prince who controls the moon's journey. With him are 84 angels who move the moon along its orbit. Each night, they guide it 354,000 parasangs, especially on the fifteenth day of the month when it reaches its turning point in the East.

Next is Rahatiel, the prince in charge of the stars and constellations. He is assisted by seven great and powerful angels. His name, Rahatiel, comes from his task—guiding the stars as they move through the sky. Each night, he leads them 339,000 parasangs, moving them from East to West and back again. The Holy One, blessed be He, has created a special path for them—a place where the sun, moon, planets, and stars rest as they travel from West to East during the night.

Following him is Kokbiel, the prince who rules over the planets. With him are 354,000 groups of ministering angels. These powerful angels guide the planets, moving them from one city and province to another within the heavens of Raqia'.

Above them all are seventy-two princes of the heavenly kingdoms, each corresponding to one of the seventy languages spoken on Earth. These princes wear royal crowns, are dressed in robes of honor, and are wrapped in magnificent cloaks. They ride royal horses and carry scepters, displaying their great authority.

As they travel through the heavens, servants run ahead of them, announcing their arrival with grand celebration. Just as earthly rulers travel with chariots, horsemen, and great armies, so do these heavenly princes make their way through Raqia'. Their journeys are marked with majesty, splendor, songs of praise, and honor. Vast armies of angels accompany them, singing and rejoicing in their greatness, just as people do when earthly kings travel in magnificent processions."

Chapter XVII

The order of ranks of the angels and the homage received by the higher ranks from the lower ones

Rabbi Ishmael said: Metatron, the angel and Prince of the Presence, explained to me:

"The angels of the first heaven, whenever they see their prince, immediately get off their heavenly horses and bow down, pressing their faces to the ground in respect. The prince of the first heaven, when he sees the prince of the second heaven, also dismounts, removes his crown, and falls on his face in humility.

The prince of the second heaven, upon seeing the prince of the third heaven, takes off his crown and bows down in awe. Likewise, the prince of the third heaven, when he sees the prince of the fourth heaven, removes his crown and lowers himself to the ground in deep respect.

The prince of the fourth heaven, when he meets the prince of the fifth heaven, does the same—removing his crown and bowing with his face to the ground. The prince of the fifth heaven, upon seeing the prince of the sixth heaven, also removes his crown and falls on his face in reverence.

The prince of the sixth heaven, when he sees the prince of the seventh heaven, takes off his crown and bows, trembling with awe. The prince of the seventh heaven, upon meeting the seventy-two princes of the heavenly kingdoms, removes his crown and falls on his face in deep humility.

The seventy-two princes, when they approach the gatekeepers of the first hall in the highest heaven, Araboth Raqia', remove their crowns and bow low in honor.

The gatekeepers of the first hall, when they see the gatekeepers of the second hall, also take off their crowns and lower themselves to the ground. The gatekeepers of the second hall, upon meeting those of the third hall, do the same—removing their crowns and bowing in respect.

This continues at every level:

- The gatekeepers of the third hall bow before those of the fourth hall.
- The fourth hall's gatekeepers bow before those of the fifth.
- The fifth hall's gatekeepers lower themselves before the sixth.
- The sixth hall's gatekeepers fall on their faces before those of the seventh.

When the gatekeepers of the seventh hall see the four great princes—the most honored ones, appointed over the four camps of the Divine Presence—they take off their crowns and bow in complete submission.

The four great princes, when they see the highest prince—the one who leads all of heaven in song and praise—remove their crowns and lower their faces to the ground in worship.

Tag'as, the great and honored prince, when he encounters Barattiel, the mighty prince who stands three fingers high in the highest heaven, Araboth, removes his crown and bows deeply to the ground.

Barattiel, when he sees Hamon, the powerful and revered prince, who is both awe-inspiring and magnificent—so powerful that all of heaven trembles when he calls out the 'Thrice Holy'—removes his crown and falls on his face in fear and respect.

Hamon, when he meets Tutresiel, another mighty prince, takes off his crown and bows to the ground. Tutresiel, when he sees Atrugiel, the great prince, also removes his crown and lowers his face in deep reverence."

Atrugiel, the great prince, when he sees Na'aririel, another great prince, removes his crown and bows down with his face to the ground.

Na'aririel, when he meets Sasnigiel, takes off his crown and lowers himself in respect. Sasnigiel, upon seeing Zazriel, removes his crown and bows deeply to show honor. Zazriel, when he encounters Geburatiel, does the same—taking off his crown and falling on his face in humility.

Geburatiel, when he sees 'Anaphiel, removes his crown and bows low. 'Anaphiel, when he meets Ashruylu, the prince who oversees all heavenly gatherings, also removes his crown and lowers himself in submission.

Ashruylu, when he sees Callisur, the prince who reveals the secrets of the divine law, takes off his crown and bows with his face to the ground in reverence. Callisur, when he meets Zakzakiel, the prince responsible for recording Israel's deeds before the Throne of Glory, removes his crown and falls on his face in humility.

Zakzakiel, upon seeing 'Anaphiel, the prince who holds the keys to the heavenly halls, also removes his crown and bows low. Why is he called 'Anaphiel? Because his honor, brilliance, and majestic presence spread throughout all the heavenly chambers, much like the Creator's glory fills the universe, as written in Habakkuk 3:3: "His glory covered the heavens, and the earth was full of His praise." In the same way, 'Anaphiel's greatness overshadows all the splendor of Araboth, the highest heaven.

When 'Anaphiel sees Sother Ashiel, a mighty and revered prince, he removes his crown and lowers himself in awe. Why is he called Sother Ashiel? Because he is in charge of the four streams of the fiery river that flows before the Throne of Glory. Every heavenly being must have his permission to enter or leave the presence of the Divine. He controls the seals of the fiery river, and his towering height measures 7,000 myriads of parasangs. When he moves before the Divine Presence, he stirs the flames of the river and declares what is written about the deeds of the world, as described in Daniel 7:10: "The judgment was set, and the books were opened."

Sother Ashiel, when he sees Shoqed Chozi, a powerful and fearsome prince, removes his crown and bows low. Why is Shoqed Chozi given this name? Because he weighs all human actions on a balance before the Holy One, blessed be He.

When Shoqed Chozi sees Zehanpuryu, a mighty and honored prince feared by all the heavenly hosts, he takes off his crown and bows in humility. Why is Zehanpuryu called by this name? Because he has the power to command the fiery river and send it back to its source.

When Zehanpuryu sees Azbuga, a prince greatly revered and exalted among those who understand the mysteries of the Throne of Glory, he removes his crown and bows down in awe. Why is Azbuga called by this name? Because in the future, he will clothe the righteous in garments of life and wrap them in robes of light, preparing them for eternal life.

When Azbuga sees two great and powerful princes standing above him, he removes his crown and falls to the ground in deep respect. These two princes are known as Sopheriel H' the Killer and

Sopheriel H' the Lifegiver. Both are ancient, mighty, and beyond reproach.

Why is one called Sopheriel H' the Killer? Because he is responsible for the book of the dead, recording the names of those whose time has come. Why is the other called Sopheriel H' the Lifegiver? Because he oversees the book of life, where the names of those whom God grants life are written, according to His will.

You may wonder, "Since God sits on a throne, do these princes also sit while writing?" The Scriptures teach (1 Kings 22:19, 2 Chronicles 18:18): "And all the host of heaven stands by Him." This makes it clear that even the greatest heavenly beings perform their duties while standing.

But how do they write while standing? One stands on the wheels of a storm, and the other stands on the wheels of a whirlwind. They wear royal garments and are wrapped in majestic cloaks. Both wear crowns of glory. Their entire bodies are covered in eyes, and they shine as brightly as lightning. Their eyes glow like the sun at full strength, and their height stretches across all seven heavens. Their wings are as numerous as the days of the year and spread across the entire expanse of the sky.

Their lips are as wide as the gates of the East, and their tongues rise as high as the waves of the ocean. Flames pour from their mouths, and their tongues burn like torches. A sapphire stone rests on each of their heads, and on their shoulders are wheels driven by swift cherubim. One holds a fiery scroll, while the other holds a flaming pen. The scroll is 30,000 myriads of parasangs long, the pen measures 3,000 myriads, and each letter they write is 365 parasangs in size.

Chapter XVIII

Ribbiel, the Prince of the Wheels of the Merkaba, and the

Surroundings of the Merkaba. The Commotion Among the Angelic Hosts During the Qedushsha

Rabbi Ishmael said: Metatron, the angel and Prince of the Presence, explained to me:

"Above these three powerful angels, there is one prince who stands apart from all others. He is honored, noble, and glorious, feared for his strength and might. He is magnificent, crowned with greatness, exalted, and beloved. There is no other prince like him. His name is Ribbiel, the great and revered prince who stands by the divine chariot.

Why is he called Ribbiel? Because he has been given authority over the wheels of the divine chariot, and they are under his control.

How many wheels are there? There are eight in total—two in each direction. Around them are four powerful winds, each with its own name: Storm Wind, Tempest, Strong Wind, and Wind of Earthquake.

Beneath the wheels, four fiery rivers flow, one on each side. Between these rivers stand four massive clouds. These clouds are known as clouds of fire, clouds of lamps, clouds of coal, and clouds of brimstone. Positioned around the wheels, they create a scene of overwhelming power and energy.

The feet of the heavenly creatures rest upon these wheels. Between each wheel, the sound of roaring earthquakes and crashing thunder echoes through the heavens.

When the moment comes for the great Song to be sung, the wheels begin to move, and the clouds shake.

At that time, all of heaven trembles:

- The mighty leaders become afraid.
- The horsemen grow restless.
- The warriors are shaken.

- The heavenly armies are filled with fear.
- The ranks of angels are overwhelmed.
- The appointed ones rush away in alarm.
- The commanders and soldiers are filled with dread.
- The servants grow weak.
- Every angel and heavenly division trembles in awe.

As the wheels turn, they call out to one another. One crown speaks to another, one heavenly creature calls to the next, and one Seraph reaches out to another, saying, as it is written in Psalm 68:3:

'Praise Him who rides upon the heavens, by His name Yah, and rejoice before Him!'"

Chapter XIX

Rabbi Ishmael said: Metatron, the angel and Prince of the Presence, explained to me:

"Above all these stands one great and powerful prince. His name is Chayy'liel. He is noble and honored, full of strength and glory. He is so mighty that all the heavenly beings tremble before him. His power is so great that he could swallow the entire earth in a single moment, as if with one bite.

Why is he called Chayy'liel? Because he has been placed in charge of the Holy Chayyoth. He strikes them with lashes of fire to stir them into action, and when they sing praises, he honors them. He urges them to proclaim, 'Holy, holy, holy,' and 'Blessed be the glory of the Lord from His place!' during the great hymn of praise."

Chapter XX

The Chayyoth

Rabbi Ishmael said: Metatron, the angel and Prince of the

Presence, explained to me:

"There are four great heavenly beings, called the Chayyoth, each connected to one of the four winds. Every one of them is as vast as the entire world. Each has four faces, and all their faces look toward the East.

Each Chayyâ has four enormous wings, each as large as the roof of the universe. Their faces contain even more faces within them, and their wings have layers upon layers of wings. The size of their faces is equal to 248 faces, and their wings are as massive as 363 wings combined.

Each of these beings wears 2,000 crowns on its head. Every crown is as beautiful as a rainbow in the sky and shines as brightly as the sun. Sparks of light radiate from them, glowing like the morning star, the planet Venus, as it rises in the East."

Chapter XXI (A)

Kerubiel, the Prince of the Kerubim, and the Description of the Kerubim

Rabbi Ishmael said: Metatron, the angel and Prince of the Presence, explained to me:

"Above all these stands one great and powerful prince. His name is Kerubiel, a noble and honored leader, full of strength and glory. His power and majesty surpass all others. He is lifted high, a righteous and holy prince, praised in every way.

Thousands of hosts celebrate him, and tens of thousands of armies exalt him. When he is angry, the earth trembles. When he is filled with wrath, the heavens shake. His presence alone causes the foundations of creation to quake, and at his rebuke, even the highest heavens tremble.

His entire body glows with burning coals. His height stretches across the seven heavens, his width spans them completely, and his form is as vast as the entirety of the heavens themselves.

When he opens his mouth, it shines like a fiery lamp, and his tongue blazes like a consuming fire. His eyebrows flash like lightning, his eyes glow with dazzling sparks, and his face burns like a raging fire.

On his head, he wears a crown of holiness, engraved with the sacred Name, from which lightning flashes forth. A divine bow rests between his shoulders. At his waist is a mighty sword, and arrows of blazing fire are strapped to his side. A shield of consuming flames hangs around his neck, surrounded by glowing coals, which encircle him completely.

The brilliance of the Divine Presence shines from his face. Majestic horns rise from his wheels, and a royal diadem rests upon his head.

His entire body is covered in countless eyes, and great wings stretch across his towering form. A burning flame rises on his right, while a fire glows on his left, with coals constantly smoldering. Firebrands erupt from his body, lightning flashes from his face, thunder echoes around him, and earthquakes rumble at his side.

Two powerful princes of the divine chariot remain beside him at all times.

Why is he called Kerubiel? Because he is in charge of the chariot of the Kerubim, the mighty beings under his command. He decorates their crowns and polishes the diadems on their heads.

He enhances their beauty and strengthens their glory. He increases their honor and leads their praise. He magnifies their splendor and refines their radiant majesty. He arranges their songs of worship to prepare a dwelling place for the One who sits upon the Kerubim.

The Kerubim stand beside the Holy Chayyoth, their wings raised high, reaching the tops of their heads. The Divine Presence rests upon them, and the brilliance of Glory shines on their faces. Songs of praise flow from their mouths, their hands are hidden beneath their wings, and their feet are covered as well. Majestic horns rise from their heads, and the light of the Divine Presence radiates from their faces.

Around them are sapphire stones and columns of fire, burning on all four sides. Fiery pillars stand beside them. A sapphire rests on one side and another on the opposite side, with glowing coals beneath them.

In every direction, divine signs stand, while their wings interlock above their heads in a display of majesty. They spread their wings in praise of the One who rides upon the clouds and glorify the awe-inspiring King of Kings.

Kerubiel, the prince who rules over them, arranges them in perfect harmony. He lifts them to new heights of honor and splendor. He strengthens them with power and beauty so they can fulfill the will of their Creator at all times. Above their exalted heads, the glory of the High King, who dwells upon the Kerubim, shines without end."

Chapter XXI (B)

Rabbi Ishmael said: Metatron, the angel, the Prince of the Presence, explained to me:

There is a great court before the Throne of Glory.

How do the angels stand in the heights of heaven? He said: They stand like a vast and endless bridge, stretching as if over a great river.

No seraph or angel is assigned to rule over it; all may approach, but the distance to cross it is enormous, measuring 30,000 parasangs. Other bridges, just as massive, extend across countless parasangs, from one end to the other. As it is written in Isaiah 6:2: "And the Seraphim stand above it and proclaim a song before Him."

The last word of this verse has a numerical value of 86, which matches the sacred name YHWH, the God of Israel. Standing before His Throne are powerful and fearsome beings—thousands upon thousands, and ten thousand times ten thousand. They lift their voices in songs of praise and worship before YHWH, the God of Israel.

The sacred text also reveals the number of bridges that exist in the heavens. There are many different kinds: bridges of fire, bridges of hail, and also rivers of hail. There are treasuries of snow and spinning wheels of fire.

There are 24 myriads of these fiery wheels. The ministering angels number 12,000 myriads—6,000 myriads above and 6,000 myriads below. In the same way, there are 12,000 rivers of hail and 12,000 treasuries of snow, equally divided—6,000 above and 6,000 below. The 24 myriads of fiery wheels are also split evenly, 12 above and 12 below. They surround the bridges, the rivers of fire, and the rivers of hail. Many ministering angels create paths through them, guiding all who stand in their midst, aligning their power with the roads of the heavens.

What does YHWH, the God of Israel, the King of Glory, do? The Great and Fearsome One, mighty in strength, hides His face.

In the highest heaven, Araboth, there are 660,000 myriads of glorious angels standing before the Throne of Glory, surrounded by blazing divisions of fire. The King of Glory covers His face, for if He did not, the heavens themselves would be torn apart by the overwhelming brilliance, majesty, beauty, and holiness of the Holy One, blessed be He.

Countless ministering angels carry out His will. There are unnumbered kings and rulers in Araboth, dwelling in His divine presence. These are honored angels among the leaders of the heavens, singing praises and remembering love. They tremble in awe of the splendor of the Divine Presence, their eyes blinded by the shining beauty of their King. Their faces darken, and their strength fades before Him.

From His Throne, rivers of joy pour forth—streams of gladness, rivers of celebration, streams of triumph, rivers of love, and currents of friendship. These waters grow stronger as they flow, surging through the gates of Araboth, carrying a mighty sound. They are accompanied by the voices of the Chayyoth marching and calling out, by the joyful tambourines of the Ophannim, and by the ringing cymbals of the Kerubim.

As these rivers swell, they rise with the song:

"HOLY, HOLY, HOLY IS THE LORD OF HOSTS; THE WHOLE EARTH IS FILLED WITH HIS GLORY."

Chapter XXI (C)

Rabbi Ishmael said: Metatron, the Prince of the Presence, explained to me:

The distance between each bridge is 12 myriads of parasangs. The path leading up spans 12 myriads of parasangs, and the path going down is the same.

The space between the rivers of dread and the rivers of fear is 22 myriads of parasangs. The distance between the rivers of hail and the rivers of darkness is 33 myriads. From the chambers of lightning to the clouds of compassion, it is 42 myriads. The clouds of compassion are 84 myriads away from the divine chariot. From the divine chariot to the Kerubim, the distance is 148 myriads. Between the Kerubim and the Ophannim, there are 24 myriads, and from the Ophannim to the inner chambers, the distance is also 24 myriads. From the inner chambers to the Holy Chayyoth, the span is 100,000 myriads of parasangs.

The space between the wings of the Chayyoth measures 12 myriads, and the width of each wing is the same. The distance between the Holy Chayyoth and the Throne of Glory stretches 130,000 myriads of parasangs.

From the base of the Throne to its seat, there are 40,000 myriads of parasangs. And the name of the One who sits upon it—let His name be made holy!

The arches of the Bow rise high above the heavens, reaching 1,000 thousands and 10,000 times ten thousands of parasangs in height. Their size matches the measure of the Irin and Qaddishin, the Warriors and Holy Ones. As it is written in Genesis 9:13: "My bow I have set in the cloud." It does not say "I will set," but "I have set," meaning it is already placed. These clouds surround the Throne of Glory, and as they pass by, the angels of hail are transformed into burning coals.

The fire of the divine voice descends from the Holy Chayyoth, but because of its power, the Chayyoth run to another place, fearing they might be sent forth. Yet they return quickly, afraid of judgment from the other side. This is why it is said in Ezekiel 1:14, "They run and return."

The arches of the Bow shine more brilliantly than the summer sun at its peak. They glow brighter than blazing fire, and their beauty and radiance have no equal.

Above the arches of the Bow stand the wheels of the Ophannim. Their height measures 1,000 thousands and 10,000 times ten thousands of parasangs, matching the measure of the Seraphim and the Heavenly Troops."

Chapter XXII

The winds blowing under the wings of the Kerubim

Rabbi Ishmael said: Metatron, the angel and Prince of the Presence, explained to me:

"Beneath the wings of the Kerubim, many different winds blow. Among them is the Brooding Wind, as it is written in Genesis 1:2: 'And the wind of God was brooding over the waters.'

There is also the Strong Wind, mentioned in Exodus 14:21: 'And the Lord caused the sea to go back by a strong east wind all that night.'

The East Wind is another, as written in Exodus 10:13: 'The east wind brought the locusts.'

The Wind of Quails is described in Numbers 11:31: 'And there went forth a wind from the Lord and brought quails.'

The Wind of Jealousy is also among them, as mentioned in Numbers 5:14: 'And the wind of jealousy came upon him.'

The Wind of Earthquake appears in 1 Kings 19:11: 'After that came the wind of the earthquake, but the Lord was not in the earthquake.'

The Wind of the Lord is described in Ezekiel 37:1: 'And He carried me out by the wind of the Lord and set me down.'

There is also the Evil Wind, as written in 1 Samuel 16:23: 'And the evil wind departed from him.'

Other winds include the Wind of Wisdom, Wind of

Understanding, Wind of Knowledge, and Wind of the Fear of the Lord, as Isaiah 11:2 states: 'And the wind of the Lord shall rest upon him—the wind of wisdom and understanding, the wind of counsel and might, the wind of knowledge and the fear of the Lord.'

The Wind of Rain is found in Proverbs 25:23: 'The north wind brings forth rain.'

The Wind of Lightnings is mentioned in Jeremiah 10:13 and 51:16: 'He makes lightnings for the rain and brings forth the wind out of His treasuries.'

The Wind that Breaks the Rocks is described in 1 Kings 19:11: 'The Lord passed by, and a great and strong wind tore through the mountains and shattered the rocks before the Lord.'

The Wind that Calms the Sea is found in Genesis 8:1: 'And God made a wind pass over the earth, and the waters subsided.'

The Wind of Wrath appears in Job 1:19: 'And behold, there came a great wind from the wilderness, striking the four corners of the house, and it collapsed.'

The Storm-Wind is mentioned in Psalm 148:8: 'Storm-wind, fulfilling His word.'

Satan is also connected to these winds, as Storm-Wind is sometimes linked to him. But all of these winds move only beneath the wings of the Kerubim, as Psalm 18:11 says: 'And He rode upon a cherub and flew; yes, He soared upon the wings of the wind.'

Where do these winds go? Scripture explains that they emerge from beneath the wings of the Kerubim and travel to the globe of the sun, as Ecclesiastes 1:6 states: 'The wind moves toward the south and turns toward the north; it continues its cycle, returning to where it started.'

From the sun, they flow to the rivers and seas, then move across the mountains and hills, as written in Amos 4:13: 'For behold, He who forms the mountains and creates the wind.'

From the mountains and hills, they return to the seas and rivers. From the seas and rivers, they pass over cities and lands. From the cities, they enter the Garden, and from the Garden, they reach Eden, as Genesis 3:8 says: 'Walking in the Garden in the wind of the day.'

Inside the Garden, the winds blend together, moving back and forth. They absorb the fragrance of the spices from every corner of the Garden. Then, they spread out again, carrying this pure scent. This fragrance is brought from the farthest parts of Eden to the righteous, who will one day inherit the Garden of Eden and the Tree of Life, as written in Song of Songs 4:16:

'Awake, O north wind; and come, O south wind! Blow upon my garden, that its spices may flow out. Let my beloved enter his garden and eat its precious fruits.'"

Chapter XXIII

The different chariots of the Holy One, blessed be He

Rabbi Ishmael said: Metatron, the angel and Prince of the Presence, explained to me:

"The Holy One, blessed be He, has many different kinds of chariots.

He has the Chariots of the Kerubim, as written in Psalm 18:11 and 2 Samuel 22:11: 'And He rode upon a cherub and flew.' These chariots are carried by the powerful and glorious Kerubim, who bear the divine presence and swiftly transport Him through the heavens.

He has the Chariots of Wind, as written in Psalm 18:10: 'And He flew swiftly upon the wings of the wind.' These chariots move as fast as the wind, carried by the forces of nature across all creation.

He has the Chariots of the Swift Cloud, as written in Isaiah 19:1: 'Behold, the Lord rides upon a swift cloud.' These chariots travel through the skies on clouds, representing divine majesty and swift judgment.

He has the Chariots of Clouds, as written in Exodus 19:9: 'Lo, I come unto you in a cloud.' Surrounded by clouds of glory, these chariots reveal His presence in mystery and splendor.

He has the Chariots of the Altar, as written in Amos 9:1: 'I saw the Lord standing upon the altar.' These chariots connect heaven and sacred places on earth, where He meets with His people.

He has the Chariots of Ribbotaim, as written in Psalm 68:18: 'The chariots of God are Ribbotaim, thousands upon thousands.' These chariots are surrounded by endless multitudes of angels, showing His unlimited power.

He has the Chariots of the Tent, as written in Deuteronomy 31:1: 'And the Lord appeared in the Tent in a pillar of cloud.' These chariots signify His presence in the Tent of Meeting, where He spoke with His people.

Each of these chariots serves a special purpose, displaying His majesty, strength, and presence as He moves through the heavens and interacts with creation.

He has the Chariots of the Tabernacle, as written in Leviticus: 'And the Lord spoke to him out of the tabernacle.' These chariots reveal His presence within the sacred space of the tabernacle, where He communicates His will.

He has the Chariots of the Mercy Seat, as written in Numbers: 'Then he heard the voice speaking to him from above the mercy seat.' These chariots represent the place where divine mercy and guidance are given.

He has the Chariots of Sapphire Stone, as written in Exodus: 'And under His feet was something like a pavement of sapphire stone.' These chariots shine with the brilliance of sapphire, symbolizing purity and heavenly majesty.

He has the Chariots of Eagles, as written in Exodus: 'I bore you on eagles' wings.' This does not mean actual eagles but refers to the speed and strength of those who carry His presence swiftly and powerfully.

He has the Chariots of Shout, as written in Psalms: 'God has gone up with a shout.' These chariots carry the triumphant and joyful proclamation of His glory through the heavens.

He has the Chariots of Araboth, as written in Psalms: 'Extol Him who rides upon the Araboth.' These chariots exist in the highest heavens, carrying His presence through the exalted realms.

He has the Chariots of Thick Clouds, as written in Psalms: 'He makes the thick clouds His chariot.' These chariots are hidden by clouds, showing the mystery and awe of His presence.

He has the Chariots of the Chayyoth, as written in Ezekiel: 'And the Chayyoth ran and returned.' These chariots, powered by the Holy Chayyoth, only move by divine command, as the Shekinah rests above them.

He has the Chariots of Wheels, as written in Ezekiel: 'And he said: Go in between the whirling wheels.' These chariots, moved by celestial wheels, represent divine power and movement.

He has the Chariots of a Swift Cherub, as written: 'Riding on a swift cherub.' When He rides upon a cherub, He places one foot upon it, and before setting the other foot down, He looks across eighteen thousand worlds. In that single moment, He sees and understands everything within them, as written in Psalms: 'You know my sitting down and my rising up; You understand my thoughts from afar.' Nothing in creation is hidden from Him.

Think of how much He perceives in just that instant, between one step and the next. When He rides upon the cherub and descends, His glory fills all creation. Then He rises again to His place. Everything exists because of His word, as written in Psalms: 'By the word of the Lord, the heavens were made.'

Who can count the number of His chariots? Scripture teaches that all these chariots stand before Him, ready to serve. There are countless myriads upon myriads of them.

The One who rides upon them knows their exact number. He has the power to set His foot upon them, and they bow before Him in complete submission, as written in Psalms: 'If I ascend into heaven, You are there; if I make my bed in the depths, You are there.'

When He rides upon the Kerubim, they carry Him. He moves upon their wings and dwells among them, yet they remain His servants, fulfilling His will.

All these chariots are bound by a divine oath to serve Him. With a single glance, He sees all of creation and rules over it. The chariots move at His command, without hesitation.

When He calls them, they respond instantly. They hear His voice, rush to carry out His will, and glorify the name of the Holy One, blessed be He."

Chapter XXIV

'Ophanniel, the Triune of the 'Ophannim

Rabbi Ishmael said: Metatron, the angel and Prince of the Presence, explained to me:

"Above all these beings, there is a powerful and exalted prince, ancient and mighty. His name is 'Ophanniel.

He has sixteen faces—four on each side—and one hundred wings on each side of his body. His form is covered with 8,466 eyes, matching the number of days in the year.

Each of his two front-facing eyes flashes with lightning, and from them, flames erupt. No creature can look directly at his eyes, because anyone who tries is instantly consumed by their intense fire.

His height is so vast that it would take 2,500 years to travel its full distance. No one can fully comprehend his size, and no words can describe his great strength—only the King of kings, the Holy One, blessed be He, knows the full extent of his power.

Why is he called 'Ophanniel? Because he is in charge of the 'Ophannim, and they have been placed under his care. Every day, he stands to serve them. He enhances their beauty, organizes their chambers, polishes their foundations, refines their dwellings, smooths their edges, and cleanses their seats. He tends to them day and night, making sure they shine with splendor, stand with dignity, and are always prepared to offer praise to their Creator.

The 'Ophannim are covered in countless eyes, and their radiance shines in every direction. On the right side of their garments, seventy-two sapphire stones are placed, and another seventy-two sapphire stones decorate the left side.

Each 'Ophan wears a crown set with four glowing carbuncle stones. These stones shine in all four directions of the highest heaven, just as the light of the sun spreads across the entire universe. Why are they called carbuncle stones? Because their brilliance resembles flashing lightning.

The 'Ophannim are surrounded by magnificent tents made of

sapphire and carbuncle, glowing with splendor and brilliance. These tents shield them and intensify the dazzling beauty of their shining eyes, filling the space around them with an unmatched, radiant glow."

Chapter XXV

Seraphiel, the Prince of the Seraphim

Rabbi Ishmael said: Metatron, the angel and Prince of the Presence, explained to me:

"Above all these beings stands a powerful and magnificent prince. He is extraordinary in every way—great, noble, honored, mighty, and awe-inspiring. He leads the heavenly hosts, moves with incredible speed, and is a scribe of unmatched skill. He is glorified, deeply respected, and loved by all.

His entire being radiates with splendor, shining with light and brilliance. Every part of him reflects beauty and greatness.

His face glows like that of angels, but his body is shaped like a mighty eagle.

His radiance flashes like lightning, his appearance burns like fire, and his beauty shines like the brightest stars. His glory glows like burning coals, his majesty sparkles like polished metal, and his brilliance resembles the glow of the planet Venus. His form reflects the light of the sun, and his height reaches the full span of the seven heavens. The light from his eyebrows is seven times brighter than normal light.

A massive sapphire stone rests on his head, as large as the entire universe, shining as brilliantly as the heavens themselves.

His entire body is covered with countless eyes, as numerous as the stars in the sky. Each eye shines like Venus, though some glow like the moon and others like the sun. Different parts of his body radiate different types of light:

- From his ankles to his knees, his glow is like flashing stars.
- From his knees to his thighs, it shines like Venus.
- From his thighs to his waist, it mirrors the brightness of the moon.
- From his waist to his neck, it radiates like the sun.
- From his neck to his head, it shines with a light that never fades.

The crown on his head is as brilliant as the Throne of Glory itself. Its size covers a distance that would take 200 years to travel. Upon this crown rests every kind of radiance, glow, and brilliance found in the universe.

This prince's name is Seraphiel, and the crown he wears is called the Prince of Peace. He is named Seraphiel because he is in charge of the Seraphim, the fiery beings under his care. Day and night, he watches over them, teaches them songs of praise, and guides them in exalting the beauty, power, and majesty of the King. He helps them sanctify His name with the greatest reverence.

The Seraphim are four in number, representing the four winds of the world. Each one has six wings, symbolizing the six days of Creation. They each have four faces.

Their size is beyond imagination—each Seraph is as tall as the seven heavens combined. Each wing is as wide as the entire sky, and each face is as vast as the whole eastern horizon.

The Seraphim shine with a light so intense that it rivals the brightness of the Throne of Glory. Their glow is so overwhelming that even the Holy Chayyoth, the mighty Ophannim, and the majestic Kerubim cannot look at them. Anyone who dares to gaze at them is instantly blinded by their incredible brilliance.

They are called Seraphim because they burn (saraph) the writing tables of Satan. Each day, Satan, along with Sammael, the Prince of Rome, and Dubbiel, the Prince of Persia, writes down the sins of Israel on tablets. These records are handed over to the Seraphim to present before the Holy One, blessed be He, in an attempt to accuse Israel.

But the Seraphim, knowing the hidden will of the Holy One, blessed be He, understand that He does not desire Israel's destruction. So what do they do? Every day, they take Satan's records and burn them in the flames that blaze before the Throne of Glory. By doing this, they ensure that these accusations never reach the Holy One, especially when He sits upon the Throne of Judgment to judge the world in truth."

Chapter XXVI

Radweriel, the Keeper of the Book of Records

Rabbi Ishmael said: Metatron, the Angel of the Lord and Prince of the Presence, explained to me:

"Above the Seraphim stands a prince of incredible greatness, higher than all other princes and more wondrous than any other heavenly being. His name is Radweriel, and he is responsible for guarding the treasuries of the sacred books.

Radweriel's main duty is to bring forth the Case of Writings, which holds the Book of Records. He retrieves this case and presents it before the Holy One, blessed be He. Once there, he breaks its seals, carefully opens it, and takes out the books inside. These sacred writings are then placed before the Holy One, blessed be He.

The Holy One, blessed be He, receives the books from Radweriel's hands and gives them to the heavenly scribes, whose job is to read them. This takes place in the Great Beth Din, the divine court in the highest heaven, in front of the entire heavenly assembly.

Why is he called Radweriel? His name reflects a unique and amazing ability—every word that comes from his mouth creates an angel. These angels are formed by the power of his voice and immediately join the ranks of the ministering angels, becoming part of the heavenly choirs.

When the time comes for the Thrice Holy song to be sung, Radweriel takes his place among the singing angels. He lifts his voice in praise, joining the celestial hosts as they glorify their Creator together."

Chapter XXVII

Description of a class of angels

Rabbi Ishmael said: Metatron, the Angel, the Prince of the Presence, explained to me:

Each angel is given seventy names, matching the seventy languages spoken throughout the world. All of these names come from the holy and exalted name of the Holy One, blessed be He. Every name is engraved in fire on the Fearful Crown, which rests upon the head of the high and glorious King.

From each name inscribed on this crown, sparks and flashes of lightning burst forth, filling the heavens with brilliant light. Surrounding each angel are majestic horns of splendor, forming a magnificent display around them. From these horns, streams of light shine outward, creating an endless glow.

Each angel is wrapped in tents of shimmering brightness, their light so intense and overwhelming that even the powerful Seraphim and mighty Chayyoth—greater than all other heavenly beings—cannot look at them. The radiance surrounding these angels is beyond imagination, a reflection of the infinite majesty of the Holy One, blessed be He, from whom their light and strength flow.

Chapter XXVIII

The 'Irin and Qaddishin

Rabbi Ishmael said: Metatron, the Angel, the Prince of the Presence, explained to me:

Above all the heavenly beings stand four great princes, known as the 'Irin and Qaddishin. These princes are highly honored, deeply respected, beloved, and full of glory. They are greater than all other heavenly beings, and no other celestial rulers or servants can compare to them. Each of these four princes is as powerful as all the others combined.

Their dwelling place is directly across from the Throne of Glory, and they stand before the Holy One, blessed be He. The brilliance of their presence reflects the light of the Throne of Glory, and their splendor mirrors the radiance of the Divine Presence.

They are glorified by the majesty of the Divine and praised in the light of the Shekina.

Not only are they deeply revered, but the Holy One, blessed be He, does nothing in the world without first consulting them. Only after seeking their guidance does He act, as it is written in Daniel 4:7: "The sentence is by the decree of the 'Irin and the demand by the word of the Qaddishin."

There are two 'Irin and two Qaddishin. How do they stand before the Holy One, blessed be He? One 'Ir stands on one side, and the other on the opposite side. Likewise, one Qaddish stands on one side, and the other on the opposite side.

These powerful princes lift up the humble and bring down the proud. They raise those who are lowly to great heights and humble the arrogant, lowering them to the dust.

Each day, when the Holy One, blessed be He, sits on the Throne of Judgment to judge the entire world, the Books of the Living and the Books of the Dead are opened before Him. All the heavenly beings stand in fear, awe, and trembling. As He sits upon the Throne of Judgment, His garments shine as white as snow, the hair on His head is as pure as wool, and His entire cloak glows with radiant light. His righteousness covers Him like armor.

The 'Irin and Qaddishin stand before Him like court officers before a judge. They bring cases forward, debate the issues, and bring each matter to a close before the Holy One, blessed be He, as it is written in Daniel 4:17: "The sentence is by the decree of the 'Irin and the demand by the word of the Qaddishin."

Some of them present arguments, while others issue rulings in the Great Beth Din in the highest heaven. Some request decisions from the Divine Majesty, while others finalize the cases presented before the Most High. Others descend to earth to carry out the judgments that have been declared, as it is written in Daniel 4:13-14:

"Behold, an 'Ir and a Qaddish came down from heaven and cried aloud, saying: Cut down the tree and remove its branches, shake off its leaves, and scatter its fruit. Let the animals flee from beneath it, and the birds from its branches."

Why are they called 'Irin and Qaddishin? Because they purify both body and spirit with fiery discipline on the third day of judgment, as it is written in Hosea 6:2:

"After two days, He will revive us. On the third day, He will raise us up, and we shall live before Him."

Chapter XXIX

The 72 Princes of Kingdoms and the Prince of the World officiating at the Great Sanhedrin in heaven

Rabbi Ishmael said: Metatron, the angel and Prince of the Presence, explained to me:

Whenever the Great Court gathers in the highest heavens, no one in the world is allowed to speak—except for a select group of extraordinary princes who have the honor of carrying the name of the Holy One, blessed be He.

How many of these princes are there? There are seventy-two, each one representing a different kingdom on earth. Above them all stands the Prince of the World, who speaks on behalf of all creation. Every day, this great Prince pleads for the world before the Holy One, blessed be He.

This takes place at the sacred moment when the Book of Records is opened. In this book, every action in the world is written down, ready for judgment. As it is written in Daniel 7:10: "The judgment was set, and the books were opened."

At this solemn time, the seventy-two princes and the Prince of the World stand before the Holy One, offering their petitions and arguments. Their purpose is to seek balance and mercy, ensuring that divine justice is carried out with compassion in the highest court of heaven.

Chapter XXX

The attributes of Justice, Mercy, and Truth by the throne of judgment

Rabbi Ishmael said: Metatron, the angel and Prince of the Presence, explained to me:

When the Holy One, blessed be He, sits on the Throne of Judgment to make decisions, three powerful forces stand around Him.

Justice is on His right, representing fairness and doing what is right. Mercy is on His left, shining with kindness and compassion. Truth stands directly in front of Him, glowing with honesty and purity.

When a person comes before Him for judgment, something remarkable happens. A staff of light emerges from the glow of Mercy and moves in front of the person. This staff is a sign that compassion is present, even in the face of judgment.

At that moment, the person falls to the ground in humility and awe. The angels of destruction, who are there to carry out punishments, tremble with fear and cannot approach because the power of Mercy is too strong.

As it is written in Isaiah 16:5: "With mercy, the throne will be established, and He will sit upon it in truth." This verse shows how justice and compassion work together on the Throne of the Holy One, blessed be He. Mercy softens judgment, ensuring that decisions are fair and true while also offering hope and redemption to those who come before Him.

Chapter XXXI

The execution of judgement on the wicked. God's sword

Rabbi Ishmael said: Metatron, the angel and Prince of the Presence, told me:

When the Holy One, blessed be He, opens the Book of Records—a divine book made of fire and flame—His judgment begins. From His presence, commands are constantly sent out, ensuring that justice is carried out against the wicked. This is done through His sword, which is drawn from its sheath.

The sword shines as brightly as lightning, and its brilliance spreads across the entire world, lighting up everything from one end of the earth to the other. As it is written in Isaiah: "For by fire, the Lord will judge, and by His sword, all flesh."

The sight of this sword fills everyone on earth with fear and trembling. Its sharp blade flashes like lightning, stretching across the horizon. Sparks and bursts of light, as bright as the stars, shoot from the sword, adding to the overwhelming power it displays. As it is written in Deuteronomy: "If I sharpen the lightning of My sword."

Chapter XXXII

When the Holy One, blessed be He, sits on the Throne of Judgment, different groups of angels take their places around Him. The angels of Mercy stand on His right, showing compassion and speaking on behalf of those being judged. On His left are the angels of Peace, radiating calm and tranquility. Directly in front of Him are the angels of Destruction, ready to carry out His commands as He sees fit.

Beneath the Throne of Glory, a scribe records everything happening in the heavenly court. Another scribe stands above the Throne, writing down the divine decrees spoken by the Holy One, blessed be He.

Surrounding the Throne on all four sides are the Seraphim, powerful beings made of fire. Their presence forms walls of light and flames that enclose the divine seat. The Ophannim also encircle the Throne, their fiery bodies covered in flames that shine in all directions. To the right and left of the Throne, there are clouds of fire, adding to the incredible majesty of the scene.

Beneath the Throne, the Holy Chayyoth carry it, each with three enormous fingers. The size of each finger is immense, measuring 800,000 and 70 times 100, plus 66,000 parasangs. These mighty beings stand in a breathtaking display of divine power, their strength holding up the Throne of Glory.

Under the feet of the Chayyoth, seven fiery rivers flow endlessly. Each river stretches 3,500 thousand parasangs wide and plunges 248 thousand myriads of parasangs deep. Their length is beyond comprehension, impossible to measure. These rivers curve and flow in all four directions of Araboth, their power spreading across the heavens.

From Araboth, the rivers descend to Mâ'ân, where they pause before moving on to Zebul, then to Shechagim, then to Raqia', and finally to Shamayim. From there, the rivers pour down upon the heads of the wicked in Gehenna, delivering divine judgment. As it is written in Jeremiah 33:19:

"Behold, a whirlwind of the Lord, even His fury, has gone forth, a whirling tempest; it shall burst upon the head of the wicked."

Chapter XXXIII

Rabbi Ishmael said: Metatron, the angel and Prince of the Presence, explained to me:

The feet of the Chayyoth are surrounded by seven layers of burning coals, each glowing with intense heat. Beyond these coals are seven walls of fire, their flames flickering and lighting up everything around them.

Outside these fiery walls are seven layers of hailstones, known as the stones of 'Elgabish, as described in Ezekiel. These hailstones shine with an icy glow, creating a sharp contrast to the fire within. Surrounding them is another layer of hailstones, called the stones of Bârâd, adding yet another powerful barrier.

Beyond these stones are layers of stormy winds, known as the wings of the tempest, swirling with uncontrollable energy. Further out, layers of flames roar fiercely, consuming everything in their path.

These flames are surrounded by the chambers of the whirlwind, constantly spinning and churning. Beyond these chambers are the realms of fire and water, two opposing elements that exist together in perfect balance, sustained by divine power.

Encircling the realms of fire and water are angels who never stop proclaiming "Holy!" as they lift their voices in praise. Beyond them are those who chant "Blessed!" in perfect harmony, their voices echoing throughout the heavens.

Surrounding the singers of "Blessed!" are glowing clouds, radiant with divine light. These clouds are enclosed by burning juniper coals, their heat intense and unrelenting. Around the juniper coals are a thousand camps of fire and ten thousand hosts of flames, each burning with incredible brightness.

Between each camp and host lies a protective cloud, shielding the heavenly beings from the overwhelming fire. These clouds act as a divine barrier, allowing the angels to fulfill their sacred duties and continue their endless worship of the Holy One, blessed be He.

Chapter XXXIV

Rabbi Ishmael said: The angel Metatron, the Prince of the Presence, explained to me:

In the highest heaven, called Araboth Raqia', God created an enormous number of angels—506,000 groups in total. Each group has 49,000 angels, and they are incredibly powerful and awe-inspiring.

Each angel is as vast as the ocean. Their faces shine as bright as lightning, their eyes glow like fire, and their arms and legs gleam like polished metal. When they speak, their voices boom like a mighty roar, full of strength and majesty.

All of these angels stand before God's Throne of Glory in four enormous rows. Each row is led by powerful angelic commanders who guide them with authority and dedication.

Some angels spend their time calling out "Holy!" while others say "Blessed!" Some act as swift messengers, rushing to carry out God's commands, while others remain still, standing in deep respect before Him. As it says in the Book of Daniel:

"Thousands upon thousands served Him, and ten thousand times ten thousand stood before Him. The court was seated, and the books were opened."

When it is time to declare the Kedushsha—the sacred praise of God—a great whirlwind suddenly bursts forth from before Him. This storm moves through the Camp of Shekhinah, shaking the heavens, just as it is written in Jeremiah:

"Look! The whirlwind of the Lord goes out in fury, a raging storm."

In that moment, thousands of angels transform into flashes of fire, streaks of lightning, and bursts of energy. Some turn into flames, rushing winds, or blazing fires. Others take the shape of glowing figures, shining like living sparks of light.

These changes happen because they fully submit to God's power. They tremble with awe and fear, overwhelmed by His presence.

They stay in this state of intense motion and energy until they completely embrace their purpose—to praise the glorious King. Once they do, they return to their original forms, standing strong and devoted. Their focus never shifts from singing praises to God, just as it says in Isaiah:

"And one called to another and said: Holy, Holy, Holy is the Lord of Hosts; the whole earth is filled with His glory."

Chapter XXXV

The angels bathe in the fiery river before reciting the Song

Rabbi Ishmael said: Metatron, the Angel, the Prince of the Presence, explained to me:

When the angels get ready to sing their Song of Praise, a powerful river of fire, called Nehar diNur, rises up, glowing with intense energy. This blazing river is filled with countless angels, each radiating strength and divine fire. It flows beneath God's Throne of Glory, passing between the angels' camps and the vast ranks of 'Araboth.

Before they can begin their song, the angels must first enter this fiery river. Each one fully immerses themselves in the flames, cleansing their spirit and preparing for the sacred task ahead. They dip their entire bodies into the fire, then carefully purify their tongues and mouths seven times, making sure their words are worthy of being offered to God.

Once purified, they emerge from the river and dress in shining robes of Machage Samal, glowing with purity and beauty. Over these, they wrap themselves in shimmering cloaks of chashmal, radiating divine light. Clothed in this sacred attire, they take their places in four perfect rows before the Throne of Glory.

This holy gathering stretches across all the heavens, with every angel standing ready to sing. In their purified state, they reflect the greatness and holiness of their Creator, prepared to lift their voices in perfect harmony to praise Him.

Chapter XXXVI

The four camps of Shekina and their surroundings

Rabbi Ishmael said: The angel Metatron, the Prince of the

Presence, explained to me:

Inside the seven heavenly halls, there are four magnificent chariots of Shekina, each shining with divine glory. In front of every chariot stand four great groups of Shekina's celestial army, filled with unmatched beauty and holiness. Flowing between these groups is a mighty river of fire, burning endlessly, symbolizing the eternal power of God's presence.

Surrounding the fiery rivers are glowing clouds, casting a soft, sacred light across the heavens. Between these clouds rise towering pillars of brimstone, standing strong as symbols of divine power and purity. Around each pillar, flaming wheels spin in endless motion, forming brilliant circles of fire.

Between these fiery wheels, flames blaze continuously, never fading. Within these flames are great treasuries of lightning, bursting with dazzling flashes that light up the heavens with awe. Beyond the lightning are the mighty wings of the storm wind, always moving, carrying out God's will across the universe.

Behind these powerful winds are the chambers of the storm, where the wild forces of nature gather and are held in place. Beyond these chambers lie vast realms filled with roaring winds, echoing voices, rolling thunder, and endless bursts of sparks. Layer upon layer, earthquakes rumble behind them, shaking one after another, their force revealing the unshakable power of the Holy One.

This breathtaking and intricate arrangement reflects divine order, with each element playing its role in upholding the chariots and the heavenly hosts of Shekina—eternal signs of God's infinite glory and rule.

Chapter XXXVII

The fear that befalls all the heavens at the sound of the Holy, especially the heavenly bodies, and their appeasement by the Prince of the World

Rabbi Ishmael said: The angel Metatron, the Prince of the Presence, explained to me:

Within the seven heavenly halls, there are four majestic chariots of Shekina, each glowing with divine light. In front of these chariots stand four great groups of Shekina's celestial army, filled with beauty and holiness beyond imagination. Flowing between these groups is a powerful river of fire, burning without end, representing the everlasting presence of God.

Surrounding this fiery river are bright, glowing clouds that spread a soft, sacred light. Between them rise towering pillars of brimstone, standing firm as symbols of divine strength and purity. Around each pillar, spinning wheels of fire create endless, dazzling circles of light.

Between these fiery wheels, flames burn without stopping. Inside these flames are great stores of lightning, sending out brilliant flashes that fill the heavens with awe. Beyond the lightning, the mighty wings of the storm winds move without rest, carrying out God's will across creation.

Behind these powerful winds are the chambers of the storm, where the forces of nature are gathered and held. Beyond them stretch vast realms filled with rushing winds, echoing voices, rolling thunder, and endless bursts of sparks. Layer after layer, earthquakes shake the space behind them, one trembling after another, revealing the unstoppable power of the Holy One.

This incredible and intricate structure reflects the divine order, with each part playing its role in upholding the chariots and the heavenly hosts of Shekina—eternal symbols of God's endless glory and rule.

Chapter XXXVIII

The explicit names fly off from the Throne, and all the various angelic hosts prostrate themselves before it during the Qedushsha

Rabbi Ishmael said: The angel Metatron, the Prince of the Presence, explained to me:

When the angels begin to chant "Holy," something incredible happens. The sacred names of God, written in fiery letters on the Throne of Glory, suddenly take flight. These powerful names rise into the sky like mighty eagles, each one shining with divine energy. They are carried by sixteen wings and soar together, circling around the Holy One on all sides of His divine presence.

As this sacred moment unfolds, all the angels and heavenly beings watch in awe. The radiant angels join with the fiery Servants, the powerful Ophannim, and the Kerubim of Shekina. The shining Chayyoth stand alongside the Seraphim, the 'Er'ellim, and the Mephsarim. Armies of fire, blazing with intense flames, gather in deep worship.

The holy princes, wearing crowns of glory, stand in robes that shine like royal garments. Their very beings radiate strength and splendor. In complete humility, they bow before God, lowering themselves three times in perfect harmony.

Their voices rise together, filling the heavens with a powerful declaration: "Blessed be the name of His glorious kingdom forever and ever." Their worship reflects the unity of creation, the purpose of their existence, and their deep reverence for the One who rules over all.

Chapter XXXIX

The ministering angels rewarded with crowns for properly uttering the "Holy," and consumed by fire for failing, with new ones created to take their place

Rabbi Ishmael said: The angel Metatron, the Prince of the Presence, explained to me:

When the angels chant "Holy" before God with the right order and deep respect, a moment of divine joy fills the heavens. The attendants of His Throne of Glory step forward, emerging with great happiness. Each of them carries countless crowns—thousands upon thousands, shining brightly like the planet Venus.

These crowns are given to the angels and the great heavenly princes who proclaim "Holy." Each one receives three crowns: the first for saying "Holy," the second for saying "Holy, Holy," and the third for completing the chant with "Holy, Holy, Holy is the Lord of Hosts." This act shows God's approval and pleasure, recognizing their devotion and place in the heavenly order.

However, if the angels fail to say "Holy" in the correct way or without proper focus, a blazing fire bursts from the little finger of God. This fire rushes into their ranks, splitting into 496,000 flames, each one directed at the four great camps of the angels. In a single moment, those who made mistakes are consumed by the flames, as it is written: "A fire goes before Him and burns up His enemies all around."

But immediately after, God speaks a single word, and from that word, new angels are created to take the place of those who were lost. These newly formed angels stand before the Throne of Glory and join in the never-ending song of praise, declaring "Holy" without hesitation. As it is written: "They are new every morning; great is Your faithfulness." Each day, fresh and renewed angels rise to continue the eternal worship of God.

Chapter XL

Metatron shows R. Ishmael the letters engraved on the Throne of Glory, by which all of creation was made

R. Ishmael said: Metatron, the Angel, the Prince of the Presence, spoke to me and said:

Come and see the letters that shaped the heavens and the earth— the same letters that formed the mountains and hills. These are the letters that created the seas and rivers, the trees, and every plant that grows on the land. They are the letters that brought the planets and stars into existence, setting the sun, moon, and great constellations like Orion and the Pleiades into their places, filling the sky with light.

These same letters were used to create the Throne of Glory, the spinning Wheels of the Merkaba, and everything needed to sustain the universe. Through them, wisdom, understanding, and knowledge came to be. They also formed virtues like patience, humility, and righteousness, which keep the world in balance.

As I walked beside him, he held my hand and lifted me onto his wings. He showed me these sacred letters, each one carved in flames upon the Throne of Glory. Sparks flew from them, their light spreading out and filling all the chambers of 'Araboth, shining across the highest heavens with their brilliance.

Chapter XLI

R. Ishmael said: Metatron, the Angel, the Prince of the Presence, said to me:

Come, and I will show you where amazing things happen. I will take you to the place where water is suspended high above, where fire burns inside hail but is never put out, where lightning flashes from snowy mountains, where thunder echoes in the highest heavens, where flames burn inside other flames, and where powerful voices sound through thunder and earthquakes.

As we walked, he took my hand and lifted me onto his wings, showing me all these wonders. I saw the waters held high in 'Araboth Raqia', kept in place by the power of the name YAH 'EHYEH ASHER 'EHYEH (I Am That I Am). From these waters, streams flowed down to the earth, nourishing it, just as it is written: "He waters the mountains from His chambers; the earth is satisfied with the fruit of His works."

I also saw fire and snow side by side, neither harming the other, kept in perfect balance by the power of the name 'ESHK 'OKLA (Consuming Fire), as it is written: "For the Lord your God is a consuming fire."

I watched as flashes of lightning shot out from snowy mountains, yet they did not fade, sustained by the power of the name YAH TSUR OLAMIM (The Everlasting Rock), as it is written: "For in Jah, the Lord, is an everlasting rock."

I heard roaring thunder and mighty voices rising from fiery flames, their sound never fading, held strong by the power of the name 'EL SHADDAI RABBA (The Great God Almighty), as it is written: "I am God Almighty."

I saw flames glowing brightly inside other flames, burning without being consumed, upheld by the power of the name CAD AL KES YAH (The Hand Upon the Throne of the Lord), as it is written: "For the hand is upon the Throne of the Lord."

I witnessed rivers of fire flowing alongside rivers of water, yet neither one destroyed the other. They remained in harmony through the power of the name OSEH SHALOM (Maker of Peace), as it is written: "He makes peace in His high places." For it is He who creates peace between fire and water, hail and flame, wind and cloud, earthquake and sparks.

Chapter XLII

Metatron shows R. Ishmael the abode of the unborn spirits and of the spirits of the righteous dead

R. Ishmael said: Metatron said to me:

"Come, and I will show you where the spirits of the righteous are—those who have already lived and returned, as well as those who have not yet been born."

He brought me close, took my hand, and lifted me up near the Throne of Glory, the place where the Shekina dwells. There, he revealed the Throne of Glory to me, and I saw the spirits that had lived and returned. They were soaring above the throne, in the presence of the Holy One.

Then, I reflected on a verse from Scripture and found its meaning in what is written: "For the spirit clothed itself before me, and the souls I have made" (Isaiah 57:16). The words "for the spirit clothed itself before me" refer to the spirits that were created in the chamber of creation for the righteous and have returned to God. The phrase "the souls I have made" refers to the spirits of the righteous who have not yet been born and remain in a chamber known as GUPH.

Chapter XLIII

Metatron shows R. Ishmael the abode of the wicked and the intermediate in Sheol. (vss. 1–6)

The Patriarchs pray for the deliverance of Israel (vss. 7–10)

R. Ishmael said: Metatron, the Angel, the Prince of the Presence, said to me:

"Come, and I will show you where the spirits of the wicked and those in between dwell, where they stand, and where they are taken. I will show you where the in-between spirits descend and where the wicked are cast down."

He said to me, "Two angels of destruction, Za'aphiel and Simkiel, send the spirits of the wicked to Sheol."

Simkiel is responsible for the in-between spirits, guiding and purifying them because of the great mercy of Adonai, the Prince of the Place. Za'aphiel is in charge of the wicked, driving them away from God's presence and the splendor of the Shekina. They are sent to Sheol to be punished in the fire of Gehenna, beaten with rods of burning coal.

As we walked, he took my hand and pointed out everything with his fingers.

I saw their faces—they looked human, but their bodies were like eagles. The spirits of those in between had a pale grey appearance because of their actions. Their sins left marks on them until they were cleansed by fire.

The faces of the wicked, however, were as black as the bottom of a pot, showing the depth of their evil and the wrongs they had done.

Then I saw the spirits of the Patriarchs—Abraham, Isaac, and Jacob—along with the spirits of the righteous. They had risen from their graves and ascended to the Heaven of Raqia'. Standing before God, they prayed with sorrow in their voices:

"Lord of the Universe, how much longer will You remain on Your Throne in mourning, with Your right hand held back? When will You rescue Your children and reveal Your Kingdom to the world? How much longer will You leave Your people as slaves among the nations? When will You show mercy once more? The hand with which You stretched out the heavens and the earth—when will You raise it again in compassion?"

God answered them, saying, "How can I act while the wicked continue to sin so terribly? How can I lift My mighty right hand when their wrongdoing has caused such destruction?"

At that moment, Metatron turned to me and said, "My servant, take the books and read their deeds!" I took the books and read the records of their actions. Every wicked soul had broken the Torah in every possible way, disobeying every law and commandment. As it is written, "Yes, all Israel has transgressed Your Law" (Daniel 9:11). The word was not written as toratecha but torateka, meaning they had violated every letter of the Torah, from the first (Aleph) to the last (Tav). Each letter of the Torah stood as a witness against them.

Upon hearing this, Abraham, Isaac, and Jacob wept bitterly. Then God said to them, "Abraham, My beloved; Isaac, My chosen one; Jacob, My firstborn—how can I now rescue these people from the hands of the nations?"

At that moment, Mihmael, the Prince of Israel, cried out in grief, weeping loudly, and said, "Why do You stand far off, O Lord?" (Psalm 10:1).

Chapter XLIV

Metatron shows R. Ishmael last and future events recorded on the Curtain of the Throne

R. Ishmael said: Metatron said to me:

"Come, and I will show you the Curtain of MAQOM, where the Divine Majesty is displayed. On it, every generation of the world is recorded—all their actions, past and future, until the end of time."

He took me with him, pointing things out with his fingers, like a father teaching his child to read the Torah. I saw every generation and its leaders:

- The rulers and heads of each generation
- The guides and shepherds
- The oppressors and those in power
- The protectors and guardians
- The judges and court officials
- The teachers and supporters
- The noblemen and warriors
- The elders and counselors

I saw Adam and his generation, their deeds and thoughts.

I saw Noah and his generation, their choices and actions.

I saw those who lived before the flood, their behavior and their fate.

I saw Shem and Nimrod, and the generation of the Tower of Babel, their struggles and their beliefs.

I saw Abraham, Isaac, and Ishmael—each with their own generation and their deeds.

I saw Jacob, Joseph, and the twelve tribes, their lives and their journeys.

I saw Moses and his people, Aaron, Miriam, the elders, and the leaders of Israel.

Then Metatron spoke:

"I saw Joshua and his generation, their victories and failures.

I saw the judges of Israel, their wisdom and struggles.

I saw Eli, Phinehas, and Samuel, and how they led their people.

I saw the kings of Judah and Israel, the choices they made, and their impact on history.

I saw the princes of Israel and the rulers of other nations, their ambitions and their deeds.

I saw the heads of councils in Israel and the nations, their leadership and decisions.

I saw the nobles, the judges, and the wise men—both of Israel and the other nations.

I saw the teachers of children in Israel and across the world, shaping the next generations.

I saw the prophets of Israel and the prophets of the nations, their messages and warnings.

I witnessed every war and battle that the nations fought against Israel during its time as a kingdom.

I saw the Messiah, the son of Joseph, and his generation, their actions, and their struggles against the nations.

I saw the Messiah, the son of David, and his time—his battles, his triumphs, and his hardships alongside Israel.

I saw the great wars of Gog and Magog in the days of the Messiah, and everything that God will do in those times.

I saw every leader, every generation, and every event—both in Israel and among the nations. Everything that has happened and everything that will take place until the end of time was written on the Curtain of MAQOM.

I saw it all with my own eyes. After witnessing these things, I opened my mouth to praise MAQOM, saying:

"For the King's word has power, and who may say to Him, 'What are You doing?'" (Ecclesiastes 8:4).

"O Lord, how great are Your works!" (Psalm 104:24).

Chapter XLV

The place of the stars shown to R. Ishmael

R. Ishmael said: Metatron said to me:

"Come, and I will show you where the stars rest, where they stand each night in the sky, filled with awe for MAQOM, and where they move from their places of rest."

I walked beside him as he took my hand, pointing out everything with his fingers. I saw the stars standing on sparks of blazing fire, surrounding the Chariot of the Almighty. Then Metatron clapped his hands, and the stars were set into motion. Instantly, they shot into the sky on fiery wings, scattering in all directions from the Throne of the Merkaba. As they soared, Metatron spoke the name of each one to me, fulfilling what is written: "He counts the number of the stars; He gives each one its name" (Psalm 147:4). This shows that the Holy One has given every star a unique name.

With perfect order, each star follows Haniel into the heavens, Raqia' hashamayim, to serve the world. When their task is done, they return in the same way to sing praises to the Holy One through songs and hymns, as it is written: "The heavens declare the glory of God" (Psalm 19:1).

In the future, the Holy One will make them new again, as it is written: "They are new every morning" (Lamentations 3:23). Then, they will open their mouths to sing a song of praise. What will they sing? It is written: "When I consider Your heavens..." (Psalm 8:3).

Chapter XLVI

Metatron shows R. Ishmael the spirits of the punished angels

R. Ishmael said: Metatron said to me:

"Come, and I will show you the souls of the angels and the spirits of the heavenly servants whose bodies have been burned by the fire of MAQOM (the Almighty). This fire comes from His little finger. These angels have been turned into fiery coals within the River of Fire (Nehar diNur), but their spirits and souls remain behind the Shekina, standing in eternal reverence.

Whenever the heavenly servants sing at the wrong time or in a way that was not commanded, they are consumed by the fire of their Creator. A flame sent by God burns them up within the chambers of the whirlwind. This mighty wind sweeps them away, casting them into the River of Fire, where they are transformed into great mountains of burning coal. However, their spirits and souls always return to their Creator, continuing to stand behind Him in devotion.

I walked beside Metatron as he took my hand and led me to see these spirits and souls. He showed me where they stood behind the Shekina, resting on the wings of the whirlwind and surrounded by walls of fire.

Then Metatron opened the gates of the fiery walls where these spirits remained behind the Shekina. I looked up and saw them. They had the forms of angels, and their wings were like those of birds, but made entirely of flames, created from burning fire.

In that moment, I opened my mouth and praised MAQOM, saying: 'How great are Your works, O Lord!' (Psalm 92:3)."

Chapter XLVII (A)

Metatron shows R. Ishmael the Right Hand of the Most High, now inactive behind Him, but in the future destined to work the deliverance of Israel

R. Ishmael said: Metatron said to me:

"Come, and I will show you the Right Hand of MAQOM, which has been placed behind Him since the destruction of the Holy Temple. From it shines every kind of light and splendor, and through it, the 955 heavens were created. Even the seraphim and Ophannim are not allowed to look at it until the day of salvation arrives.

I walked beside him, and he took my hand. With joy, songs, and praise, he showed me the Right Hand of MAQOM. No words can fully describe its beauty, and no eyes can withstand its greatness, majesty, and glory.

By its side stand the souls of the righteous, those found worthy to witness the joy of Jerusalem. They praise and pray before it three times a day, saying, "Awake, awake, put on strength, O arm of the Lord." As it is written: "He caused His glorious arm to go at the right hand of Moses."

At that moment, the Right Hand of MAQOM was weeping. From its five fingers, five rivers of tears flowed down into the great sea, shaking the entire world. This is as written in Scripture: "The earth is utterly broken, the earth is torn apart, the earth shakes violently. The earth will stagger like a drunken man and sway like a hut." These five events are connected to the five fingers of His mighty hand.

When the Holy One sees that there is no righteous person left in the generation, no one devoted to goodness, and no justice among people—when there is no one like Moses or Samuel to stand before Him and pray for salvation and deliverance—when no one calls upon His Right Hand to act on behalf of Israel, then He remembers His own justice, mercy, and grace. By His own power, He will bring salvation. As it is written:

"He saw that there was no one, and He was amazed that there was no one to intercede; so His own arm brought salvation, and His righteousness upheld Him."

Scripture reminds us how Moses constantly prayed for Israel in the wilderness, preventing divine judgment, and how Samuel called upon God, who answered his prayers, even when they were not part of the divine plan. As it is written: "Moses and Aaron were among His priests" and "Even if Moses and Samuel stood before Me."

At that time, the Holy One will say: "How long shall I wait for humanity to bring salvation through their righteousness? For My own sake, for My merit and justice, I will stretch out My arm and redeem My children from among the nations." As it is written: "For My own sake I will do it, for how can My name be defiled?"

Then the Holy One will reveal His mighty arm to the nations. It will stretch across the whole world, shining with the brilliance of the summer sun at its peak.

In that moment, Israel will be saved from the nations. The Messiah will appear and lead them to Jerusalem with great joy. They will feast and celebrate, glorifying the Messianic Kingdom and the house of David throughout the world. No nation will have power over them anymore. The people of Israel will gather from the four corners of the earth and dine with the Messiah. But the nations of the world will not share in their feast, as it is written:

"The Lord has bared His holy arm in the sight of all the nations, and all the ends of the earth shall see the salvation of our God."

And again: "The Lord alone led him, and there was no foreign god with him."

"And the Lord shall be King over all the earth."

Chapter XLVII (B)

The Divine Flames that go forth from the Throne of Glory, crowned and escorted by numerous angelic hosts through the heavens and back again to the Throne—the angels sing the Holy and the Blessed

Metatron said to me:

These are the seventy-two sacred names that are written upon the heart of the Holy One. They are names of power, righteousness, and majesty: SeDeQ, SaHPeL, SUR, SaDdiQ, SeBa'oTh (Lord of Hosts), ShaDdaY (God Almighty), 'eLoHIM (God), YHWH, and many others of great holiness. Among them are names like ROKeB 'aRaBOTh (He who rides upon the Araboth), HaY (The Living One), and QQQ (Holy, Holy, Holy). Each of these names carries deep meaning and mystery, declaring His eternal glory and dominion. They affirm His strength and wisdom, as written:

"He gives power to the weary and increases strength to those who have no might."

These names are surrounded by countless crowns—crowns of fire, crowns of flame, crowns of chashmal, and crowns of lightning. They are accompanied by thousands upon thousands of powerful angels, carrying them with honor like subjects escorting a mighty king. These angelic hosts surround them with pillars of fire, glowing clouds, flashes of lightning, and brilliant light. Wherever they move, there is awe, trembling, majesty, and deep reverence, along with dignity, glory, wisdom, and understanding. Their journey is marked by the brightness of chashmal and the splendor of divine radiance.

As they travel through the heavens, these sacred names are praised. The angels call out before them, "Holy, Holy, Holy!" The heavenly hosts roll them through the realms of the skies, treating them as honored and mighty princes.

When these names are finally brought back to the Throne of Glory, the Chayyoth surrounding the Merkaba open their mouths in praise. They declare the holiness and greatness of His name, saying:

"Blessed be the Name of His glorious kingdom forever and ever."

Chapter XLVII (C)

An EnochMetatron piece

I took him, strengthened him, and gave him a special purpose. I chose Enoch, my servant, who is unlike any other among the children of heaven. I made him strong during the time of the first Adam. But when I saw how corrupt the people of the flood generation had become, I removed my Shekina from among them. I lifted it up to the heavens with the sound of a trumpet and a mighty shout, as it is written:

"God has gone up with a shout, the Lord with the sound of a trumpet."

I took Enoch, the son of Jared, from among humans and raised him up to the high heavens with the sound of a trumpet and a loud cry. I made him my witness among the Chayyoth of the Merkaba in the world to come. I gave him a throne that stands near my own Throne of Glory, measuring seventy thousand parasangs, all made of fire. I assigned him seventy angels, representing the seventy nations of the world, and gave him authority over all realms, both in heaven and on earth.

I granted him wisdom and understanding greater than all other angels. I gave him the name The Lesser Yah, a name whose Gematria value is seventy-one. I placed him in charge of the works of creation, making his power greater than that of the ministering angels.

I appointed him over all the treasuries and storage places in every heaven and gave him the keys to each one. He became the prince over all heavenly rulers, a minister of the Throne of Glory, and the overseer of the Halls of Araboth, with the authority to open their doors before me. He was given the responsibility to arrange and exalt the Throne of Glory.

I placed him over the Holy Chayyoth, crowning them with honor, and over the majestic Ophannim, strengthening them with glory. He was assigned to clothe the exalted Kerubim with majesty and make the radiant sparks shine with brilliance.

I gave him authority over the flaming Seraphim, covering them with greatness, and over the Chashmallim, filling them with radiant light. His task was to prepare my seat every morning. As the highest prince, he ensured that the Holy Chayyoth were crowned with majesty and clothed with honor, ready to carry out their divine roles.

I seated him before my Throne of Glory so that he could magnify my name in its fullness. He was entrusted with revealing my power and holding the secrets of both the heavens and the earth. He was chosen to witness my greatness as I sat upon my throne in majesty and splendor.

I made him greater than all others, raising him to a height of seventy thousand parasangs among the mighty. His throne was exalted to reflect my own, and its glory was increased to match the honor of my presence.

I transformed his body—his flesh became blazing fire, and his bones turned into burning coals. His eyes shone like flashes of lightning, and his eyebrows glowed with endless light. His face radiated like the sun, and his eyes reflected the majesty of the Throne of Glory.

I dressed him in honor and majesty, wrapping him in beauty and greatness. On his head, I placed a royal crown—a diadem of unmatched brilliance, measuring five hundred by five hundred parasangs. I adorned him with my own honor, majesty, and the splendor that shines from my Throne of Glory.

I called him The Lesser YHWH, The Prince of the Presence, and The Knower of Secrets. I revealed every mystery to him, just as a father shares knowledge with his son. I entrusted him with all secrets, ensuring that he would proclaim them in righteousness and truth.

I established his throne at the entrance to my Hall, where he sits to judge the heavenly hosts. Every prince of heaven stands before him, receiving instructions from him to carry out the will of the Most High.

Seventy names, taken from my own names, were given to him to elevate his status. I placed seventy princes under his command, ensuring they followed my words in every language.

I gave him the power to humble the proud and raise up the lowly. With a single word, he could bring down kings and redirect their paths. He was granted the authority to establish rulers in their positions, as it is written:

"He changes the times and the seasons; He removes kings and sets up kings." (Daniel 2:21)

He was tasked with giving wisdom to the wise and knowledge to those who seek understanding, as it is written:

"And He gives knowledge to those who understand." (Daniel 2:21)

He was chosen to reveal the secrets of my words and teach the laws of my righteous judgment, as it is written:

"So shall my word be that goes forth from my mouth; it shall not return to me empty but shall accomplish what I desire." (Isaiah 55:11)

The phrase "I shall accomplish" is not used, but rather "he shall accomplish", meaning that whatever command comes from the Holy One, Metatron carries it out faithfully. He upholds and establishes the decrees of the Holy One.

I entrusted him with teaching the Law, the Books of Wisdom, the Haggada, and the Tradition, ensuring that those who study them gain complete understanding. As it is written:

"Whom will He teach knowledge? And whom will He make understand tradition? Those weaned from milk, taken from the breast." (Isaiah 28:9)

Chapter XLVII (D)

Metatron has seventy names, which the Holy One took from His own name and gave to him to increase his glory. These names include Yehoel Yah, Yehoel, Yophiel, Aphphiel, Margziel, Simkam, Yahseyah, Ssbibyah, Periel, Tatriel, Tabkiel, and many others. Each name reflects the divine power and holiness given to him. One of these names is The Lesser YHWH, because God's own name is within him, as it is written:

"For My name is in him." (Exodus 23:21)

Another name, Sagnesakiel, refers to his role as the guardian of all the treasuries of wisdom.

All the knowledge of wisdom was entrusted to Metatron, and it was through him that this wisdom was revealed to Moses on Mount Sinai. During the forty days Moses stayed on the mountain, he learned the Torah in seventy forms and seventy languages. He also studied the Prophets, Writings, Halakhas (laws), Traditions, Haggadas, and Toseftas, all in seventy forms and languages. This covered every aspect of divine knowledge and law. However, at the end of the forty days, Moses forgot everything he had learned in an instant.

Then the Holy One called Yephiphyah, the Prince of the Law, and through him, all the knowledge was restored to Moses as a gift. As it is written:

"And the Lord gave them unto me." (Deuteronomy 10:4)

From that moment on, the knowledge remained with Moses forever. And how do we know this? It is written:

"Remember the Law of Moses, My servant, which I commanded him at Horeb for all Israel—My statutes and judgments." (Malachi 4:4)

Here, "the Law of Moses" refers to the Torah, Prophets, and Writings. "Statutes" refer to the Halakhas and Traditions, and "Judgments" refer to the Haggadas and Toseftas. All of this sacred knowledge and wisdom was given to Moses directly from the heavens on Mount Sinai.

The seventy names given to Metatron reflect the Explicit Names engraved on the Merkaba and the Throne of Glory. These sacred names were taken from God's own names and placed upon Metatron. The ministering angels use these seventy names to address the King of Kings in the highest heavens. Alongside these names, there are also twenty-two letters engraved on the ring of His finger. This ring is used to seal the destinies of the heavenly rulers, the Angel of Death, and the fate of every nation on earth.

Metatron is known as the Angel, the Prince of the Presence, and the Prince of Wisdom, Understanding, Kings, Rulers, and Glory. He was honored above all the great beings of heaven and earth. He himself testified, saying:

"The God of Israel is my witness that when I revealed this great secret to Moses, all the heavenly hosts rose up against me in outrage."

The angels demanded to know why such sacred knowledge—the very secret by which the heavens, the earth, the seas, the mountains, the rivers, Gehenna, the Garden of Eden, the Tree of Life, and even the Torah itself were created—was given to a mortal man.

"Why would you share this with someone born of a woman, a being made of flesh, imperfect and unclean? Did you get permission from the heavens? Were you granted authority from the Holy Place?"

In response, Metatron declared that the Holy One had indeed given him the authority and permission to reveal these secrets. But the angels were still not satisfied until God Himself stepped in. He rebuked them and said:

"I delight in Metatron, my servant. I have chosen him, loved him, and entrusted him with these mysteries. He is unique among all the children of heaven."

With God's approval, Metatron took these treasures of wisdom and revealed them to Moses. Moses then passed them on to Joshua, who gave them to the elders. The elders then passed them down to the prophets, and from the prophets, they were entrusted to the men of the Great Synagogue.

From there, they were handed to Ezra the Scribe, then to Hillel the Elder, and later to R. Abbahu, who passed them to R. Zera. R. Zera then entrusted them to the men of faith, whose role was to use this knowledge to guide, warn, and heal all the diseases afflicting the world.

As it is written:

"If you listen carefully to the voice of the Lord your God, do what is right in His eyes, pay attention to His commands, and keep all His laws, I will not bring upon you any of the diseases I brought upon Egypt, for I am the Lord who heals you." (Exodus 15:26)

And so, this sacred wisdom was passed down through the generations, preserving the knowledge and healing power entrusted by the Creator of the World.

(Ended and finished. Praise be to the Creator of the World.)

Thank You for Reading

Dear Reader,

We hope this timeless classic has sparked your imagination and enriched your literary journey. Now that you've turned the final page, we want to share a vision for the future of reading—one where every classic you've ever wanted to explore is at your fingertips, in a format that best suits your life.

We'd like to invite you to gain immediate, unlimited digital & audiobook access to hundreds of the most treasured literary classics ever written—along with the option to secure deluxe paperback, hardcover & box set editions at printing cost. Together, we can spark a new global literary renaissance alongside our small, independent publishing house called "The Library of Alexandria."

Thousands of years ago, the Library of Alexandria stood as a beacon of knowledge—until it was lost to history. We aim to reignite that spirit of preservation and discovery right now, in the modern age—only this time, it's accessible to all, in every language and every format.

Picture a world where every timeless classic, novel, poem, or philosophical treatise is not only available to read but also updated for today's readers—modernized, translated into any language or dialect, and ready to enjoy in any format you choose, whether that is in an eBook, audiobook, paperback, or deluxe hardcover & box set version a printing cost.

By joining our movement to rebuild the modern Library of Alexandria, you become part of an unprecedented mission to offer:

- **Unlimited Audiobook & eBook Access to the Greatest Classics of All Time**

  Instantly explore thousands of legendary works, from Plato and Shakespeare to Jane Austen and Leo Tolstoy. All are instantly ready to read or listen to, giving you a complete literary universe at your fingertips.

- **Paperback & Deluxe Editions at Printing Costs:**

  Purchase any title in a paperback, deluxe hardbound, or deluxe boxset edition at printing costs, shipped right to your doorstep. Curate your personal library of Alexandria with editions worthy of display—crafted to last, designed to captivate, and delivered straight to your door.

- **Modern translations for Contemporary Readers in all languages and dialects**

  Discover a vast selection of classics reimagined in clear, current language—no more struggling with outdated phrases or obscure references. Next to the original versions, we aim to offer translations in as many languages and dialects as possible.

  As we continue our translation efforts and add new languages, readers everywhere can connect with these works as if they were written today. By bridging linguistic divides, you're contributing to ensuring that these timeless stories become more meaningful, accessible, and inspiring for people across the globe.

- **Your Personal Library of Alexandria:**

  Over the months and years, you'll curate a unique physical archive of classics—each volume a testament to your taste, curiosity, and love of knowledge. It's not just about owning books—it's about curating a cultural legacy you'll cherish and pass down for generations to come.

- **Join a Global Literary Renaissance:**

  Your support fuels an ongoing mission: allowing us to reinvest in offering deluxe print editions (including special boxsets) at their true cost, broaden the range of available formats and translations, and extend the reach of these works to new audiences worldwide. By joining today, you're not just preserving a legacy of masterpieces; you set in motion a powerful wave of literary accessibility.

  We are more than a publisher—we're a movement, and we can't do it alone. Your support lets us scale our mission, preserving and reimagining history's greatest works for tomorrow's readers.

Become a Torchbearer of knowledge.

Thank you for picking up this book and allowing us into your literary journey. As you turn the pages, know that you're part of something larger: a global effort to keep these stories alive, share their wisdom across borders and generations, and spark a true cultural revival for the modern era.

If this resonates with you—please consider taking the next step by visiting:

www.libraryofalexandria.com

With gratitude and a shared love of knowledge,

The Modern Library of Alexandria Team

Visit:

www.libraryofalexandria.com

Or scan the code below: